RADICAL KNOWLEDGE:

a philosophical inquiry into the nature and limits of science

RADICAL KNOWLEDGE:

a philosophical inquiry
into the nature and limits of science

by GONZALO MUNÉVAR

with a Foreword by PAUL K. FEYERABEND

HACKETT PUBLISHING COMPANY
1981

To Pat and Ryan

First printing
Printed in the United States of America
Copublished in the United Kingdom by
Avebury Publishing Company Ltd

For further information, please address
Hackett Publishing Company, Inc.
P.O. Box 55573
Indianapolis, Indiana 46205

Library of Congress Cataloging in Publication Data

Munévar, Gonzalo.
Radical knowledge.

Bibliography: p.
Includes index.
1. Science—Philosophy. I. Title.
Q175.M9813 1981 501 81-4258
ISBN 0-915145-17-0 AACR2
ISBN 0-915145-16-2 (pbk.)

CONTENTS

PREFACE AND ACKNOWLEDGMENTS

I do not remember exactly when I first came by the ideas in this essay, but I think they began to crystalize during my first year of graduate school at Berkeley, in conversations with my friend and neighbor Mike Williams. That was in 1971, when many of our fellow students still went around town breaking windows in support of some cause or other. In the midst of all that excitement Mike and I would sit instead and talk about the nature of knowledge, interrupted, it is fair to say, by games of Go and other pleasant activities – and on one occasion by the fire-bombing of an automobile in front of my apartment. It is perhaps inappropriate that from such sedate mentality should have arisen an essay by the title of *Radical Knowledge*, although I would like to think that the passion of my contemporaries was not absent in me but simply directed elsewhere.

After a brief preview on Benson Mates' blackboard, I gave the first public presentation of my central theses on radical epistemology at one of Paul Feyerabend's seminars during my second year at Berkeley. I had feared that Feyerabend would tear me to pieces, but he invited me instead to lecture to his graduate class the following term. For several months after that he treated me to lunch and conversation at the outdoor section of the Golden Bear, a setting that provided a wonderful opportunity to sharpen my views on many subjects and to bring *Radical Knowledge* to maturity. Paul Feyerabend is probably the most interesting person I have ever met, and his friendship has been very valuable to me over the years; but his devastating criticism is the sort one would wish on one's worst enemy, or on oneself when taking seriously the notion that criticism is at the heart of progress and improvement of ideas. The man questions everything in the most disconcerting fashion, even obvious claims come up for challenge and sometimes ridicule. Indeed, when we discussed my first draft I had to fight him every line of the way. I do not remember whether we ever agreed on anything, but I am still certain that this work owes much to those stimulating conversations over lunch. His writings perhaps did not have as much of an influence on my thought, except for the habit (which I may have also picked up from Lakatos) of consigning a lot of material to the footnotes. I hope the reader will not find disagreeable the marked preference in this essay to carry only the main argument in the text and make the appropriate historical and scientific remarks on the side.

During my years as a graduate student (and since) I was also very fortunate to benefit from Michael Scriven's insightful commentary, and from the kind direction of Gunther Stent, who gave me so generously of his time that I feared his laboratory might to unattended.

In the years I have been a teacher of philosophy *Radical Knowledge* has evolved rather slowly, partly because of my new obligations, and partly because I wanted to take into consideration the commentary of the many people whose care and interest were bestowed upon earlier versions of the manuscript. I thank Cliff Anderberg, Roy Edgley, Nancey Murphy Fedan, Michael Gillespie, Ernst Mayr, Wallace I. Matson, Hans Sluga, and E. O. Wilson. I owe a special debt of gratitude to Ralph Berger, David Paulsen, and Sheldon Reaven, whose dedication to my manuscript has not lagged behind my own, even though their views and mine may often be at odds. I am sure there must be others I have forgotten to mention; to them I now offer my thanks and my apologies.

I am very grateful to the American Council of Learned Societies for the grant that made the final draft possible. I am also grateful to David Lamb, the editor of this series, Jim Dening of Avebury Publishing Co., Bill Hackett of Hackett Publishing Co., and to Beverly Walker for her fine typing.

It is quite nice, in putting the finishing touches to this manuscript, to be able to enjoy the love and support of my wife, Patricia, as well as the curiosity of my infant son, Ryan, who often insists in giving me a hand at the typewriter.

FOREWORD

In this concise, clear and beautifully written essay Dr. Gonzalo Munévar restates the problem of knowledge and reality and solves it in a simple and ingenious way. Realizing that philosphical positions contain assumptions about the world he criticizes the separation of science and philosophy. Using scientific results, especially results in biology and the physiology of perception, he argues for a scientific relativism: knowledge comes from interactions between an organism and its surroundings, and perceptions and theories are 'relative' to frames of reference in precisely the sense in which mass, distance and other physical magnitudes are relative to frameworks in the theory of relativity. Replacing a purely conceptual account of knowledge by a performance view he can discuss the limitations of knowledge in biological terms, explain total knowledge (approximately) as a theory that exploits all the resources of the genotype in question and make the flexibility of a theory an important part of its excellence.

The epistemology that emerges is elucidated with the help of examples, compared with familiar views such as those of Mach, Spencer, Popper, Kuhn, Lakatos and Toulmin, and defended against simple(minded) as well as against sophisticated objections. Biology, history and the performance model of knowledge play a large role in this defence, but they are not used in a dogmatic manner: some historical examples occur in their standard interpretation as well as in alternative interpretations and arguments is provided for both.

A most attractive feature of the book is the absence of useless precision and sophistication. This does not mean that technical arguments are avoided. It means that the author has always tried to state, and almost always succeeded in stating, the underlying idea in a non-technical way. The philosopher will learn that Dr. Munévar's position, while lacking important features of traditional epistemologies, does not lead to disorder and irrationality – but rationality is here part of science and not an outside agency. The scientist will welcome the book as a comprehensible piece of philosophy that can aid him in his research. The educated layman will get a better insight into the complex nature of our knowledge than most philosophy-of-science textbooks can provide. Using this insight both expert and layman may be better able to cope with the intellectual challenges that lie ahead.

Paul Feyerabend
Meilen 1981

1

ON THE NATURE OF PHILOSOPHY: PRELIMINARY REMARKS

Philosophy is the 'love of wisdom', as one can well read in any introductory text. With the exception of some cold-hearted individuals, everybody knows what love is. But wisdom? That is a much harder thing to determine. Wisdom is somehow associated with conduct, with being able to do the right thing.[1] At one time in our Western philosophical past wisdom was also associated with knowledge. This book is in part an attempt to restore that association while righting many wrongs of philosophical method and substance. This aim is neither too ambitious nor too lofty, for its fulfillment demands only that we recognize the truly radical nature of scientific knowledge.

Throughout much of our history, philosophical systems were supposed to give us a complete picture through an investigation of the nature of reality. Systems were rational, or teleological; everything had a purpose, a function in reality. Thus philosophy investigated the nature of reality. It was presumed that knowledge of the nature of reality would give us knowledge of the function of man (the 'meaning of life'), and hopefully procedures for applying such findings to individual cases. In those days, religion, ethics, epistemology, metaphysics, and science were closely interwoven. The rise of modern science, triggered by the Copernican revolution, changed the philosophical landscape. When we no longer explained the physical world in terms of reasons and purposes, the teleological obsession about reality lost ground, withered, and finally died as a rational position. Religion was the main casualty of the change of outlook.

Philosophy has not remained untouched: it has separated from science and has had great difficulty coming to terms with itself. Since Kant's failure to deal with Hume's embarrassment of empiricism, great efforts have been made to carve out a niche where professional philosophers can pursue a course apart from that of the sciences while at the same time preserving their own intellectual respectability. Hume argued that, if all knowledge comes from experience, the sense of order in nature that results from applying the so-called scientific method is unwarranted, for in science we generalize from examined cases to unexamined ones (of which we have no experience). For example, we do not experience causality, hence our talk about causal laws is in need of justification. Kant tried to provide the justification by proposing that the rational mind has organizing principles (categories) with which it gives order to its experience of the universe. Such categories constitute the structure of the mind and are required for the very possibility of empirical knowledge. The categories are then 'prior' to experience. But there are some problems with Kant's categories. Are they appropriate (justified)? Why these and not others? Where did they come from? Kant had no satisfactory answer to the first question. The second became practically unanswerable with the downfall of Newtonian physics (on which Kant had based some of his categories). The third one has seldom been considered (except by the early evolutionary epistemologists, who, as we will see in chapter 6, had some extremely interesting answers to the other questions as well).[2]

In a neo-Kantian fashion Western philosophy has become an *a prioristic* discipline. Given that philosophy studies what is logically prior to experience (e.g., concepts), the practice of it must be *distinct* from that of the sciences, in objectives as well as in methodology. Such a separation has come to be considered fundamental: the progress of science has been said to be impeded by philosophical elements (the Vienna Circle); the intrusion of scientific models into philosophy leads only to conceptual confusion and shows a lack of sensitivity to the philosophical enterprise (Wittgenstein); science can only give us objective knowledge, but such knowledge is already a finished product, passive, and thus ultimately irrelevant to the human condition (Kierkegaard).

Ever since the time of the Royal Society and Newton there has been a dinimution of speculation and metaphysical bias in science. But such a clear break between science and philosophy is a rather contemporary phenomenon. Science has increasingly come to be regarded as an axiomatic enterprise, at least ideally.[3] Such a static view of knowledge, of scientific knowledge in particular, has resulted in another apparent separation: between knowledge and wisdom (between facts and values, between facts and decision).

What of the 'love of wisdom'? Don't we still want to be wise? As I mentioned before, being wise is somehow associated with behavior, or rather with a program for behaving in some special way. It seems that we are in the confines of ethics. But 'behaving' should not be taken in the narrow sense of behaving towards others (that is, of social behavior). It should be taken in a larger sense: what to do with respect to oneself, to other humans, to other creatures, to man's creation, to man's world, solar system, universe. But to do so one needs knowledge, knowledge of what one has to deal with. And what does man have to deal with, now and in the future? It seems that again we can indulge in an old-fashioned style and say that we need knowledge of the nature of reality. Such is one of the connections between knowledge and wisdom. The relationship between knowing and doing things is closer still, however, as I will explain later on. So is the relationship between science and philosophy.

I have come to believe that there is a much more intimate connection between science and philosophy than is generally admitted, particularly by analytic philosophers. In fact, I find a quasi-dialectical relationship between the two. I think that a parallel can be made to the relationship that exists between science and technology (in that science could be thought to determine the boundaries within which technology may operate, and even suggest the avenues of endeavor most profitable for technology to pursue), the advancement of technology enables us to view and test our science in ways not open to us before. As a result we can change our science, which leads to the possibility of further advancing our technology, and so on. Philosophy, epistemology and metaphysics in particular, can and should be tested by science in an analogous way. The task does not seem so far-fetched if one realizes that philosophers make crucial, and often implausible, empirical assumptions in what they think are pure conceptual investigations. With some luck the result of 'setting philosophy straight' will also benefit science.

Such closeness between knowledge and action, and philosophy and science can be better seen after a careful examination of the nature and limits of scientific knowledge. This task will be carried out in four steps as follows.

(1) The biological foundation of knowledge leads to the *relativity* of perception, intelligence, and science.
(2) The proper model of knowledge is one of *performance.*
(3) The limits of knowledge cannot be determined by conceptual or methodological means.
(4) Science is a rational enterprise; but we can realize this only if we accept a *social conception* of scientific rationality.

Together, these four steps will lead to a radical epistemology that looks forward, to *radical knowledge.*

(1) It will be seen that perception and intelligence are the result of a special kind of interaction between an organism and the environment. The sorts of experience an organism has are determined to a large extent by the perceptual apparatus it possesses and by the environment in which it finds itself. This view will be developed so it can be shown that (1) in analyzing the relationship between theory and observation, philosophers must make crucial empirical assumptions, and (2) those empirical assumptions are often implausible.

If one must make empirical assumptions it is sensible to make plausible ones. Doing so leads to adverse consequences for two important positions: (1) that observation can be used as a neutral standard by which the merits of competing theories can be evaluated, and (2) that the method of making hypotheses, testing them, and replacing them gets us closer and closer to the 'structure of reality' (the so-called hypothetical realism of Campbell, Lorenz, and Popper).[4] Both positions, the first in particular, have been the subject of great controversy in the past few years. That controversy will be alluded to but not elaborated upon in these pages. My methodological considerations provide a different approach to this matter. The position that results is a new kind of relativity which sheds new light on, and perhaps even solves, the most crucial problems in the philosophy of nature. Relativism will thus no longer be the philosophical pariah it has remained for thousands of years.

(2) It will be argued that scientific knowledge is roughly a species of understanding, and that the criterion for understanding is one of performance, of action (a parallel will be made to Michael Scriven's comprehension criteria for information processing devices). Knowledge is then viewed as the ability to interact with the environment in certain ways.

An interactionist view of perception, intelligence, and scientific knowledge forces us to consider the matter of evolution. Just as evolution is relevant to systems interacting with the environment at the biological level and, as Konrad Lorenz and E. O. Wilson have shown, at the behavioral level, we are faced with the issue of the relevance of neo-Darwinism to epistemological matters. Herbert Spencer, Ernst Mach, and to some extent Henri Pioncaré developed an evolutionary theory of knowledge in the late nineteenth century. At its heart was the thesis that just as the present interaction of a biological system with its environment is at least partly a result of a long evolutionary process, the present intellectual mechanisms which enable us to interact with our environment in certain ways are also the result of a long evolutionary process. Their views were plagued, however, with crudely formulated and often unwarranted connections between scientific merit and survival value. Mach, in particular, could be thought to have claimed that specific scientific theories are products of a historical succession of biological processes.

In recent years Karl Popper and Stephen Toulmin have also proposed an evolutionary epistemology. Unlike their nineteenth century counterparts, their positions sever all connections with survival value, and may be called biological only by analogy. (Toulmin, for example, would want to treat the history of science in an evolutionary manner, with the best theories being selected because they adapt better to the intellectual environment in which they struggle with their competitors; insofar as the history of science has remained faithful to such a schema it should be considered rational.)

I will claim that whereas the nineteenth century evolutionary epistemology was too strong, the contemporary one is too weak. My own version may well serve as a compromise. I will sketch it in chapters 4 and 5. In chapter 6 I will contrast it with those of the aforementioned authors, and will also comment on some of Lorenz's contributions. At this point I may emphasize an important difference with approaches such as Toulmin's. Whereas he, properly impressed by neo-Darwinism, decides to extrapolate evolutionary theory to other fields, and particularly to the epistemology of science, I proceed differently: starting from an examination of the nature of scientific knowledge I am led to a situation in which I must face the relevance of evolutionary concerns to my views. In a manner that should have pleased Lakatos, my evolutionary epistemology is a 'natural' consequence of my basic program.

(3) The examination of basic intellectual structures in an evolutionary context will be very useful in the investigation of the limits

of scientific knowledge. This third step will be carried out in chapters 7 and 8, where two main lines of argument will be pursued. One will be that no purely conceptual or methodological determination can be made of the limits or the manner of growth of scientific knowledge. We are not in a position to make such a determination, we may never be, and we will require science itself to make any strides in this area (a rather strange claim about an issue that sounds so philosophical). Of special interest will be two popular notions: that science must develop within the confines of two-value logic, and that science must preserve what has been confirmed by prior (and respectable) scientific practice.

The other line of argument deals with a rather touchy problem. Some people fear that any attempt to bring evolution into an examination of man will result in a justification of the *status quo* (whether epistemological, moral, or social). It is true that evolution has been misused for precisely that purpose. But my investigation leads me to the opposite conclusion.[5] Thus *Radical Knowledge* will stand for an epistemological view that is interactionistic and relativistic, and that advocates not only the possibility of the overthrow of conceptual establishments but the desirability of attempting it.

(4) The picture that emerges of scientific knowledge is a dynamic one. I agree with Jean Piaget when he says:

> Scientific knowledge is in perpetual evolution; it finds itself changed from one day to the next. As a result, we cannot say that on the one hand there is the history of knowledge, and on the other its current state today, as if its current state were somehow definitive or even stable. The current state of knowledge is a moment in history, changing just as rapidly as the state of knowledge in the past has ever changed and, in many instances, more rapidly. Scientific thought, then, is not momentary; it is not a static instance; it is a process.[6]

A dynamic epistemology leads to a dynamic view of rationality. Since change is part and parcel of scientific knowledge, the emphasis should not be on a search for reasons or for common, neutral standards (for comparing theories).[7] Scientific rationality, I will argue, does not lie on the particular system of beliefs that individual scientists hold, neither at the beginning nor at the end of any change; nor does it lie *exclusively* in the manner in which they commit themselves to it. Furthermore, a social conception of rationality is in order because rationality turns out to be a social (or structural) property of the scientific enterprise as a whole.

As a result of a social and dynamic view of rationality (mainly in chapter 4) we might escape the dilemma posed to the traditional account of the growth of science (viz. that science is a cumulative enterprise) by some objections raised in the past few years. The traditional view has come into great difficulty when challenged by Kuhn and Feyerabend's claim that changes of theory involve changes of 'world views' (and by Feyerabend's case against method). The problem arises because changes of world views also involve changes in the evaluative standards, and without common standards of evaluation there does not seem to be a way to show the rationality of the scientific enterprise. This is the famous issue of incommensurability. At one time Feyerabend seemed to think that incommensurability need not prevent us from claiming that conceptual change has brought about a better theory, thus allowing us to show the rationality of the change. But recently he has come to agree with his early critics, although, unlike them, he is quite happy with the results. That he has come around can be seen from the following:

> Now I do not see how the desirability of revolutions can be established by Kuhn. Revolutions bring about a *change* of paradigm. But following Kuhn's account of this change, or 'gestalt-switch' as he calls it, it is impossible to say that they have led to something *better*. It is impossible to say this because pre- and post-revolutionary paradigms are frequently incommensurable.[8]

It is clear then that Feyerabend shares the assumptions made throughout the history of philosophy about the procedures for establishing rationality. Elsewhere he finds it exhilarating that such procedures can never be fulfilled.[9] But since it is not on an appeal to standards that rationality should be based, but on our preparedness to change (as I will argue), we have no cause for despair (nor has Feyerabend for rejoicing).

These are the four steps by which I will show the proper relationship between philosophy and science. Because of the methodological considerations that will lead to my views, and because of the actual content of such views, this essay will not belong to the neo-positivism that characterizes so much of contemporary philosophy of science.

Now, the only challenge to the analytic tradition in the philosophy of science serious enough to be considered an alternative has been the historical approach of Kuhn, Feyerabend, Lakatos and others. Of the two approaches, I much prefer the latter, not only because it remedies the lack of attention to the scientist's actual struggles and issues (at least in the history of physics and chemistry, which it confronts with more than lip service) but because I also consider it a much more insightful way of doing philosophy.[10]

In spite of my obvious leaning toward the historical approach, this essay will not be one more contribution to it, however. To keep in mind the history of physics is fine, to be aware of the history of science as a whole is admirable. But I do not want to restrict myself to looking back. Science is not very old, even if we think that it began with the Pre-Socratics, or earlier in Egypt and Babylon. History is a very scholarly discipline. Science is far more than scholarly.[11] The essential ingredients for its practice are not restricted to intellectual tough-mindedness, they also include the opening of great possibilities and the imaginative task of making them plausible. With luck we and our science will be around for many millions of years, perhaps till our universe tires into disintegration. I want an epistemology of science that can be prepared to deal with the surprises that such a long future may offer us. Or one that can at least begin to deal with them. *Radical Knowledge* is an attempt to provide such an epistemology. It should be properly regarded as a sketch toward an alternative conception, as a beginning, not as a finished product. Not only because substantial alternatives must take a long time to develop, but because as will be seen throughout the essay, finished products should be regarded with suspicion.

References

1. The connection is ultimately with behavior, but it seems safer to talk first about conduct, for some people think of the wise man as someone who meditates much and acts little. Such a caricature may cast a paradoxical air on the claim that wisdom is associated with behavior.
2. Questions of generation are not considered philosophically legitimate nowadays, even if they are interesting on other grounds. But we will see later how answers to such questions may throw light on better certified 'philosophical' issues.
3. Russell, for example, claimed that mathematics did not become science until the axiomatization of geometry by Euclid.
4. Since all the issues alluded to in this chapter will be treated in detail in the body of the essay, I do not provide references here except in cases where the need for them is quite specific.
5. This may well be taken as a defense of sociobiology. I will have more on this topic toward the end of the book.
6. J. Piaget, *Genetic Epistemology*, Norton and Co. (1971), p. 2.
7. Someone may wonder whether my claims here could amount to a standard of sorts. Perhaps so, but not in the strict way normally involved in the contemporary discussions of the rationality of science.
8. Paul K. Feyerabend, 'Consolations for the Specialist', in I. Lakatos and A. Musgrave (eds.), *Criticism and the Growth of Knowledge*, Cambridge University Press, (1970), p. 202.
9. See, for example, his *Against Method*, NLB (1975).

10. Perhaps in a century or so neither of these two will be seen as very important representatives of our century, even though they now dominate the attention of professional philosophers. Many, if not most, great philosophical revolutions have come about as the result of the struggle to establish a new scientific discipline. So it might be that the methodological and theoretical disputes of sciences such as molecular biology and sociobiology will bear the mark of 20th Century philosophy.

11. I do not intend to downgrade history, of course. I do claim that science is very young given the total history of mankind to date, and given the scientific period itself, which I hope will extend for a very long time. This very last point will be considered in chapter 5.

2

ON THE WAY THINGS REALLY ARE: THE PROBLEM OF REALITY

In determining the nature of scientific knowledge, the first (trivial) point that comes to mind is that science is supposed to give us knowledge about the world. Whether science actually does what it is supposed to do is a famous philosophical problem. Science has been traditionally divided between observation and theory. After this distinction is made the once trivial point becomes a complicated issue: What is the relationship between theory and observation? Are they both necessary for the acquisition of knowledge? Is theory perhaps only a 'story' that allows us to order the facts so we can better keep track of them?

Facts are facts, one hears, but theories are something else altogether. No one wants to be caught denying facts, but theories have to be proven (or validated, or confirmed, or corroborated, or undergo falsification attempts, or suffer who knows what other indignities). In plainer words, theories have to be tested, sometimes by complicated experiments, sometimes by simple observations. Just what the test amounts to becomes the bread and butter of the philosophy of science. To some, if the outcome is positive the theory receives support (inductivism). To others, support is temporary, for the only interesting outcome is a negative one, in which case the theory is said to be refuted, or falsified, and the scientist must come up with a different theory (falsificationism). This is a very simplified story, of course (philosophical positions have a way of becoming very sophisticated). Nonetheless, all views of this sort have one thing in common: theories are to be judged by the way they measure up to the facts. All these methodologies (which will be discussed in more detail later) require a neutral standard of comparison between theories: facts. It is not entirely clear what facts (in the abstract) are. But the requirement is apparently fulfilled by the next best thing: careful and unbiased observations. Statements describing (or perhaps referring to) such observations comprise the so-called observation languages.

Not all views are of this sort, however. The instrumentalists, and conventionalists in general, claim that facts do not decide between theories, for with appropriate adjustments many theories can account for the facts just as well. The decision must be made on grounds of simplicity, elegance, or whatever. At any rate, the requirement of the neutrality of the observation language prevails in conventionalism as well, as I will argue shortly.

It is clear that the role of sensory experience as a stable, solid foundation of empirical knowledge is of extraordinary importance in the history of the philosophy of science. Such a point of view, however, has increasingly come under attack from thinkers such as Hanson, Feyerabend, Popper, and Kuhn. 'But is sensory experience fixed and neutral? Are theories simply man-made interpretations of given data?' Kuhn asks.[1] 'The epistemological viewpoint that has most often guided Western philosophy for three centuries dictates an immediate and unequivocal Yes!'[2] Nonetheless, he adds

> In the absence of a developed alternative, I find it impossible to relinquish entirely that viewpoint. Yet it no longer functions effectively, and the attempts to make it do so through the introduction of a neutral language of observation now seem to me hopeless.[3]

What if the distinction between theory and observation were convincingly denied? What, that is, if a good case could be made for the claim that observations were theory-laden?

The solidity of the observation language has been challenged on precisely those grounds. There are no such things as pure observations: they all require a prior point of view, a bias – either a theoretical one, or one that has come about through learning.[4] Imre Lakatos has given a very nice breakdown of the different positions on this issue:

> There is an important demarcation between *'passivist' and 'activist' theories of knowledge.* 'Passivists hold that true knowledge is Nature's

imprint on a perfectly inert mind: mental *activity* can only result in bias and distortion. The most influential passivist school is classical empiricism. 'Activists' hold that we cannot read the book of Nature without mental activity, without interpreting them in the light of our expectations or theories. Now *conservative 'activists'* hold that we are born with our basic expectations; with them we turn the world into 'our world' but must then live forever in the prison of our world. The idea that we live and die in the prison of our 'conceptual framework' was developed primarily by Kant; pessimistic Kantians thought that the real world is for ever unknowable because of this prison, while optimistic Kantians thought that God created our conceptual framework to fit the world. But *revolutionary activists* believe that conceptual frameworks can be developed and also replaced by new, *better*, ones; it is *we* who create our 'prisons' and we can also, critically, demolish them.[5]

Many think that chaos would result if the purity, the neutrality, of the observation language could not be consistently maintained (cf. the response to Kuhn and Feyerabend). Not every contemporary approach to philosophy of science indulges in such dire predictions, though. For example, Lakatos' methodology, which I will discuss in chapter 7, does not. The same can be said for *hypothetical realism.* According to this last view, science is made up of *hypotheses* about the structure of the world (of *reality*). Such hypotheses should be severely tested, and replaced by better ones (i.e. that do as much as those refuted, and then some), which in turn must be severely tested, and so on. Even observation languages can be refuted and replaced (so they may be properly called observation theories). By this process we may never arrive at the structure of reality, but we come closer and closer.

Let me summarize briefly the various traditional positions with respect to the relationship between science and reality. On the one hand we have the field of pro-scientific (or objectivist) philosophies. Some of these hold that observation languages give us descriptions that correspond to the world, and dispute as to whether, or in which form, theories do the same (e.g. empiricism and naive falsificationism). And others question the reliability of observation languages but place their hopes on a process that will presumably bring theory into closer correspondence with reality (e.g. hypothetical realism). Still others try to wash their hands of this matter of correspondence, but end up in the same boat with most empiricists (e.g. conventionalism). On the other hand we have positions which deny either that there is anything for science to correspond to, or that we are justified in believing that there is. Reality is in the mind, so to speak, according to the position that denies the first claim (subjectivism). The second constitutes the sceptical challenge that has been termed 'the problem of the external world'.

Not everyone agrees that there are two sides to this issue of reality. Much of analytic philosophy of science tried to show that there were no sides at all. It was a pseudo-problem. So the trivial point that science is about the world became unfashionable for quite some time. Why? Presumably because philosophical questions (for which there are no decision procedures) were asked as if they were empirical (for which there were supposed to be decision procedures). Confusion about the character of philosophical questions led to expectations for answers that could not be forthcoming. At the heart of this analytic approach is the thesis of the separation between science and philosophy.

I will argue that the thesis of separation is mistaken. My methodological points will put me at odds with the analytic tradition. I will then have to consider whether any objectivist position is correct. I will hold that none are. This is not to make a case for the traditional non-objectivist positions – subjectivism and scepticism. I will develop instead a relativistic position that may serve as a compromise between the two sides of the issue about reality. But such a compromise may offer no comfort to the standard philosophical mentality, for the radical epistemology at its heart neglects all previous views on the matter.

The first step in the construction of such a radical epistemology is my thesis of the relativity of perception, intelligence and science, which I discuss in the following chapter. As a perhaps necessary prelude to that thesis it is worth considering in some more detail the connection between realism and the main positions on the nature of scientific knowledge.

The first of the traditional objectivist positions – classical empiricism – claims that we get our knowledge from experience. Thus, observations become the solid basis on which scientific knowledge is grounded. It can be fairly said that the epistemology of classical empiricism, or common sense epistemology, is a passivist view (according to Lakatos' classification). Popper has called it 'the bucket theory of the mind': knowledge drips into the bucket through several holes: the senses. Such a theory assumes that empirical knowledge grows by repetition of sensory experiences.[6] But repetition, as Popper has said, presupposes similarity, and 'similarity presupposes a point of view – a theory, or an expectation'.[7] It is also clear that we could not just open our eyes and ears, cleanse our nostrils, ready our tongues, and let the facts enter our mind. If we were to proceed in such fashion empirical knowledge could not even begin. For only some facts, and some kinds of facts, will be relevant to any particular investigation. But then again, relevance presupposes a point of view.[8]

The naive position, of course, has not been held by the most sophisticated contemporary empiricists. The solidity of the observation language can find a comfortable compromise in what Lakatos called 'optimistic Kantianism'. Yes, we do have expectations, (or to paraphrase Popper, we are endowed with the analogue of a computer program that organizes our perceptions). But happily enough they lead us to an observation language that corresponds to reality, and that constitutes a solid base on which to build our science (or, if one is a falsificationist, an anvil against which to crack unworthy hypotheses and theories).[9]

Since it is not difficult to dispose of the so-called common sense epistemology, the important criticism has been aimed at the more sophisticated view. For example, it is often pointed out that there exist many interesting perceptual variances among cultures, and that perception is influenced by learning and other factors.[10] I intend to provide a general argument within which these perceptual pot-shots against the neutrality of the observation language can take their proper place. Such will be the starting point for revolutionary activism in epistemology, for *radical knowledge.* Before I do so, I will discuss two other views, one that also requires the neutrality of the observation language, and one that attacks it.

Whereas plain empiricists have normally thought that both their observations and theories correspond to reality in some sense, logical empiricsts argued that theories are only instruments that we can use to account for the observational data. Although their approach seemed quite revolutionary at the time, I think that the logical empiricists are stuck with a dependence on the solidity of the observation language, and as they themselves require it, with the neutrality of it.[11] For a very short period Carnap proposed a pragmatic theory of observation, but it is no accident that that period was very brief. At the heart of instrumentalism (and of conventionalism in general) is the notion that many different theories can 'save the phenomena' just as well. Preference among them must be a matter of simplicity, elegance, or convenience. If we were allowed radical revisions of our observation language, if we were allowed to change our observational prisons as we change our theoretical ones, then the main conventionalist tenet would be a vacuous one, for different theories would deal with not altogether like phenomena. Logical empiricists are not forced to claim the certainty of observational language, but they surely have to vouch for its reliability within the limits of observational error.[12] If they are not prepared to do so, they become pessimistic Kantians (or at least agnostics in the matter). If they take a more radical stand on observation languages they lose sight of their own position. Historically this has been the case. Neurath, for example, is accused of irrationalism by Popper and Lakatos. According to Popper:

> We need a set of rules to limit the arbitrariness of 'deleting' (or else 'accepting') a protocol sentence. Neurath fails to give any such rules and thus unwittingly throws empiricism overboard . . . Every system becomes defensible if one is allowed (as everybody is, in Neurath's view) simply to 'delete' a protocol sentence if it is inconvenient.[13]

The point is not whether all propositions are fallible, which they are, but that Neurath does not offer a rational strategy to guide us when they clash. So in Lakatos' words 'the only alternative he seems to offer [to falsificationism] is chaos'.

As I mentioned earlier, the road to epistemological radicalism begins with the critical examination of the requirement for a neutral observation language. Feyerabend, Lakatos, and Popper think of themselves as revolutionary activists. Feyerabend goes too far, all the way into epistemological anarchy.[14] Lakatos does not go far enough (cf. my arguments in chapter 7), neither does Popper. Popper's view has evolved and now can be fairly called part of the tradition of evolutionary epistemology which began with Mach. According to this view both our observational and theoretical prisons change by the pressure of evolution. Mach, Popper (and also Konrad Lorenz, who is a most able exponent of the view) are 'hypothetical' realists.[15] Not only can we overthrow and replace the theoretical parts of our science, but we can also do the same with our observational theories. Any section of our knowledge is open to revision. *But* we must replace those views with alternatives that get us closer to *the* structure of *reality*. In Popper's words:

> There is a closely related and excellent sense in which we can speak of 'scientific reality': the procedure we adopt involves (as long as it does not break down, for example because of anti-rational attitudes) success in the sense that our conjectural theories tend progressively to come nearer to the truth; that is, to true descriptions of certain facts, or aspects of reality.[16] . . . The task of science which, I have suggested, is to find satisfactory explanations, can hardly be understood if we are not realists. For a satisfactory explanation is one which is not *ad hoc*; and this idea – the *idea of independent evidence* – can hardly be understood without the idea of discovery, or progressing to deeper layers of explanation: without the idea that there is something to discuss critically.[17]

When Popper argues for realism, he is claiming that our theories must correspond to reality, at least roughly. Progress is marked by closer correspondence (this matter of correspondence will receive attention shortly).

We can realize, then, that realism, the philosophical 'paradigm' I want to attack, is presupposed by all the main positions on the nature of scientific knowledge (although at different levels: sometimes at the level of observation, sometimes at that of fundamental entities). Realism involves several metaphysical and epistemological assumptions: (1) 'things' (the universe) *are* 'out there', which is a good assumption, and that (2) they are 'out there' in one and only one *way* (the structure of reality). With these two assumptions comes a heuristic fiction (or a 'cognitive ideal'): it is assumed (3) that a creature that 'really' knows about the universe would know the way things really are. Thus, if, say, God is cosmologically omniscient, He will have complete knowledge of the way things really are. Now come the disputes. The metaphysicians dispute as to *which* way things really are;[18] the epistemologists as to whether we, poor humans, can have knowledge of such a way.

A naive realist would claim that we 'experience', e.g. see, things as they really are. More 'sophisticated' philosophers, such as some analytic philosophers, have claimed that we actually experience, e.g., see, 'representations' of the way things really are. If under 'normal' conditions we see 'green' when looking at grass, then grass *must* be green; or, perhaps, our seeing green on such occasions is *evidence* that grass is green.

Even more sophisticated should seem Popper's version of hypothetical realism and his adaptation of Tarski's conception of truth to natural languages and scientific theories. Philosophers of very different persuasions have voiced their great enthusiasm for Tarski's semantic conception of truth. Popper, for one, believes that it provides a clear and trouble-free version of the truth-as-correspondence-to-the-facts theory. I am not sure that Tarski himself would endorse this interpretation. But be that as it may, I would like to consider the epistemological ideal that goes hand in hand with such a theory of truth.

The concern seems to be the relationship between empirical knowledge in general (science in particular) and reality. Truth should hinge on reality, Quine claims. 'The sentence "Snow is white" is true, as Tarski has taught us,' he says, 'if and only if *real* snow is *really* white.'[19] (my italics) In Popper's words:

> The English statement 'Grass is green' corresponds to the facts if, and only if, grass is green.[20]
>
> We can now say that what Tarski did was to discover that in order to speak about the correspondence between a statement *S* and a fact *F*, we need a language (a meta-language) in which we can *speak about* the statement *S and state* the fact *F*. (The former we speak about by using a metalinguistic expression '*f*' which *states or describes F*.)
>
> The importance of this discovery is that it dispels all doubt about the meaningfulness of talking about the correspondence of a statement to some fact or facts.
>
> Once this is done, we can, of course, replace the words 'correspondence to the facts' by the words 'is true'.[21]

This interpretation of Tarski not only rehabilitates the idea of truth as correspondence to the facts, according to Popper,[22] but is also 'absolutistic' and 'destructive of relativism'.[23]

Now a cosmologically omniscient being, say God, would have a complete list of all true statements about the universe.[24] That would be a main difference between God and us at the present. Another difference is that God would have certainty and we do not. This is so because our senses can mislead us sometimes either by mistakes or by self-generating the experiences (as it happens to people deprived of sensory information for many hours). God of course would not have this second difficulty for he would enjoy direct acquaintance, so to speak, with the way things really are.

This assumption (3), or 'cognitive ideal', is so straightforward as to seem beyond question. Of course an omniscient being would have complete knowledge of *the* way things really are. Just as straightforward is the concomitant assumption (2): reality can be structured in only one way; to claim otherwise is to fall prey to the paradoxes of relativism (the universe is like this or like that, but not both). On the other hand, assumption (1) – 'the existence of the external world', as it is so often stated – has been made the center of philosphical controversy by the sceptic. In rallying against the sceptic's position, contemporary epistemologists have tried to show that there is something 'conceptually', 'grammatically', or 'linguistically' wrong with any questioning of assumption (1). Success on this point and Popper's success in providing a correspondence theory of truth (à la Tarski) would seem to make realism practically impregnable.[25] Nevertheless realism crumbles on closer examination.

According to Popper, psychology should be regarded as a biological discipline, and any psychological theory of the acquisition of knowledge, in particular, should be so regarded.[26] Konrad Lorenz has made this same point very nicely. And so have many other thinkers influenced by biology. At the heart of many of the resulting views is the notion that human cognition, from sensation to intelligence, is the result of an interaction (or a history of interactions) between an organism (species) and its environment. From interactionism we move on to recognize the importance of evolution to the development of human cognition (thus, evolutionary epistemology). But from such interactionism we also move on to the demise of all present forms of realism since, as we will see in the next chapter, interactionism leads to a view that is 'relativistic' and 'destructive of absolutism'.

The argument can be sketched rather simply. If psychology depends on biology, there is no reason to suppose that there can be only one kind of psychology, for many are the paths of biology. It will be seen, then, that no matter how good a perceptual or conceptual frame of reference is, many others may be just as good (there are no 'preferred' frames). As a result, the search for *the* structure of reality is comparable to the search for the absolute mass – or velocity – of an object given the principle of relativity in Einstein's Special Theory. Along these lines there will emerge a radical epistemology that can withstand the traditional objections against relativism.

References

1. T. S. Kuhn, *The Structure of Scientific Revolutions*, Prentice Hall (1970), p. 10.
2. *Ibid.*
3. *Ibid.*
4. A primitive science would presuppose commonsense beliefs. Many philosophers have resisted the idea of considering commonsense a theory. Commonsense, compared with what we think a theory should be, is more sloppy and full of hidden assumptions, many of them unwarranted. Such characteristics may make us hesitant about applying the term 'theory' to the body of commonsense beliefs, but it does not remove it from the realm of criticism.
5. I. Lakatos, 'Falsification and the Methodology of Scientific Research Programmes', in *Criticism and the Growth of Knowledge*, I. Lakatos, A. Musgrave (eds.), Cambridge University Press (1970), p. 104.
6. Partly the root of the notion that physical laws are empirical generalizations.
7. K. R. Popper, *Objective Knowledge (An Evolutionary Approach)*, Oxford University Press (1972), p. 24. This sort of view can also be traced back to Plato.
8. It does not help to say that we should restrict ourselves to those facts relevant to a problem, for problems by themselves do not distribute relevance to facts (we need a point of view to tackle the problem).
9. Philosophers' inability to 'justify' such optimism has been, in some

form or another, the central problem of epistemology for a long time. The sceptics' criticism was limited to pointing out the lack of justification, given certain assumptions that they shared with the other side. The criticism that follows in the text, on the other hand, has tried to offer an epistemological alternative. Konrad Lorenz, who was not a professional philosopher, made the interesting claim that since our brain has been attuned to the environment through evolution, we have a natural tendency to develop appropriate gestalts, or mental sets, when immersed in careful observation of that environment. There are several difficulties with this kind of proposal, however, as will be seen in chapter 6. See Konrad Lorenz, *Studies in Animal Behavior*, Harvard University Press (1971), pp. 246–53, and p. 281.

10. See, for example, N. R. Hanson, *Patterns of Discovery*, Cambridge University Press, chapter 1.

11. Only in the light of such requirement can one understand Carnap's attempt to derive physical concepts from 'erlebs'.

12. Incidentally, Duhem, famous for his conventionalism, believed in a progression of theories toward a final natural classification. Presumably such final classification would not be a matter of convention.

13. K. R. Popper, *The Logic of Scientific Discovery*, Harper & Row edition.

14. See, for example, his 'Against Method', *Minnesota Studies for the Philosophy of Science*, 4 (1970).

15. The expression was coined by D. T. Campbell, an exponent of the view.

16. K. R. Popper, *Objective Knowledge*, *op. cit.*, p. 40.

17. K. R. Popper, *ibid.*, p. 203. His qualification is worth taking into account: 'And yet it seems to me that within methodology we do not have to presuppose metaphysical realism; nor can we, I think, derive much help from it, except of an intuitive kind. For once we have been told that the aim of science is to explain, and that the most satisfactory explanation will be the one that is most severely testable and most severely tested, we know all that we need to know as methodologists'.

18. The problem of the status of theoretical entites, and ontological disputes in general, falls under this heading.

19. W. V. Quine, *Philosophy of Logic*, Prentice Hall (1970), p. 10.

20. K. R. Popper, *Objective Knowledge*, *op. cit.*, p. 315.

21. *Ibid.*, p. 316.

22. *Ibid.*, p. 308, and also p. 317.

23. *Ibid.*, p. 308.

24. This particular formulation leads to several difficulties. See next chapter.

25. As far as I can tell no one has been successful with this kind of approach.

26. K. R. Popper, *Objective Knowledge*, *op. cit.*, p. 42.

3

THE RELATIVITY OF PERCEPTION, INTELLIGENCE, AND SCIENCE

The case for *radical knowledge* starts with a simple idea: that at an elementary level the experiences of an organism are the result of an interaction between its biology and its environment. This very simple idea is connected with other simple ideas: that perception has a biological basis, that the structures of intelligence arise out of perception, and that science is a product of intelligence. Such an innocuous sounding group of ideas leads to the most surprising results when its consequences for our notions about scientific knowledge are pushed to the bitter end. Of particular significance is the discovery that philosophical positions such as realism make assumptions which are not only *empirical* but implausible. This discovery leads to two very important results: a relativistic conception of reality, and a far closer relationship between philosophy and science than analytic philosophers are prepared to admit. Both these results will become clear in this chapter as the arguments for the relativity of perception, intelligence, and science unfold in that order.

If a person is color blind he will look at the grass and will not see green but a shade of gray.[1] We say that he does not see things as they are. Now suppose a person, Man_2, sees colors which are always the complements of the colors we see; for example, when looking at grass he will see red.[2] We might again say that he does not see things as they are. But in this second case some contemporary philosophers may feel puzzled: Does Man$_2$ know the meanings of color words? A Wittgensteinian may point out that the problem is whether Man$_2$ picks out the same objects we do in circumstances in which we pick out 'green' objects. That is, is his use (application) of color words consistent with ours? If it is, then there is no issue, for how are we to tell that he sees red when we see green and vice versa? Furthermore, someone may argue that since there is no way to tell so, the case cannot be understood (what does it amount to?). On the other hand if his use is not consistent with ours we are to say that he has not learned the language, or at least that portion of it. For having learned the language consists precisely in being able to do the things he cannot do (like applying the right color words).

Nevertheless I think there can be an issue here. I will sketch several cases of interest, and then I will suggest how we could tell. First I will make the very reasonable assumption that, given a certain optical input, a creature's visual experience will depend to a large extent on the visual apparatus it is equipped with. Next I will proceed with a thought experiment. The account of perception I will provide is obviously contrived and a great oversimplification, but I think that the features essential to the matter have not been distorted.

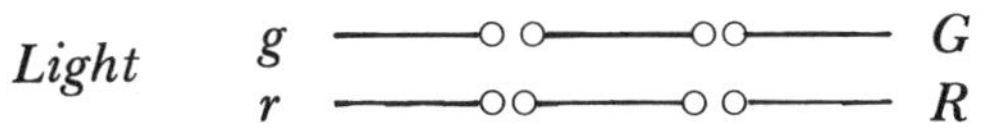

Light *g* — G
r — *R*

Green light stimulates receptor g, which will result in the brain undergoing a 'green' brain process G (a brain process correlated with 'seeing green'). Red light stimulates receptor r, and so on. Now suppose that some wires get crossed:

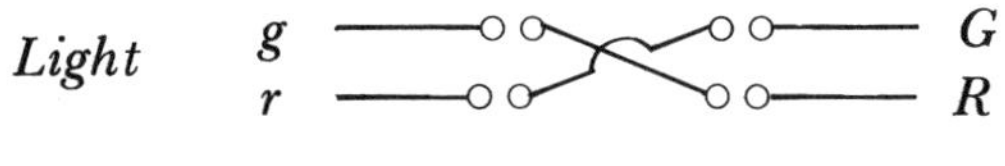

Light *g* — G
r — *R*

We can realize that Man$_2$ will see red when receiving green stimuli and vice versa. (I must repeat that this is an oversimplified case. I do not want to suggest that color perception is brain-process specific in humans, though in lobsters it is.)[3] Consider now the following cases.

Case 1: A whole species, Creatures$_1$ (C$_1$), (from Cassiopeia, say), which is biologically identical to homo sapiens in all other respects, has the green-red circuits crossed, as in the case of Man$_2$. The spectrum of radiation to which they visually respond can be correlated to their experience as follows:

	Input (measured in Angstroms)	*Experience*
	N + K A	Infrared
	N A	green
	N − K A	orange
	N − 2K A	yellow
Visual response	N − 3K A	red
	N − 4K A	blue
	N − 5K A	indigo
	N − 6K A	violet
	N − 7K A	ultraviolet

Above and below the line neither species responds. In humans the green and red are switched around.

Case 2: Another species, Creatures_2 (C_2), for which *all* the complementary colors are switched around.

Case 3: This species, Creatures_3 (C_3), responds to a wider spectrum, from N + K A to N − K A, and they experience from infrared to ultraviolet, passing through the normal order of colours for humans.

Case 4: These creatures (C_4 respond to the same range of frequencies as humans, but their experiences are moved one color all the way down the line. That is, to N A they respond with the experience of infrared. Thus violet would be just outside their range of experience. How can this be? Suppose that in humans we could divide the visual apparatus into ten stages, $S_1 - S_{10}$. Now, let us imagine that between what would correspond to the humans' S_5 and S_6 the C_4's have an additional stage $S_{5½}$, such that the equivalent reaction to a frequency step down occurs. Thus if the input was N A, by the time it came to S_6, whereas humans would have a certain reaction, for C_4's it would be as if they had received N + K A originally. Their first experience-step would be like that of ants, say, even though their first stimulus-step would be like that of humans.

Case 5: Creatures_5. Whereas in humans and $C_1 - C_4$ there was a *linear* relation (more or less), between the frequency of the input and

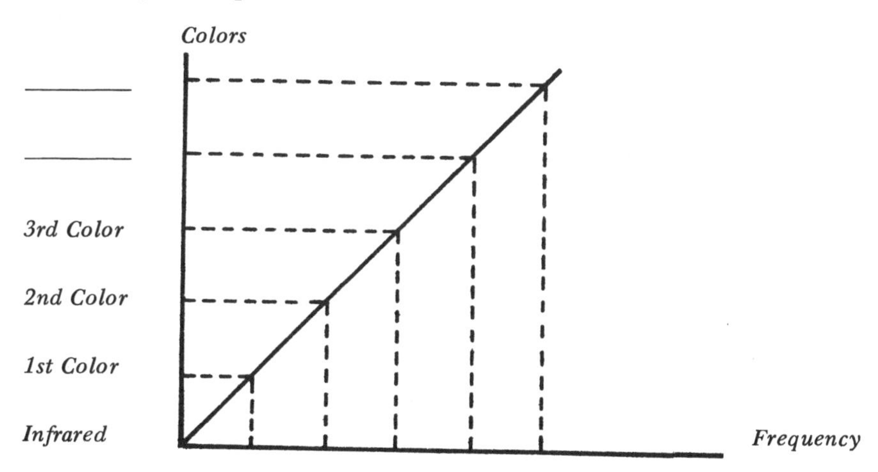

Varieties of correlation between experience and perception

the color experience, for C_5's it is an exponential relation, of some other non-linear relation.

As seen from the C_5's graph, there is not a one-to-one correspondence between their distribution of colors and those of humans and $C_1 - C_4$. In some cases where the latter would make more than one color distinction the C_5's would make just one. And as the spectrum approaches the ultraviolet the C_5's make several color distinctions where the other species make just one.

Case 6: Creatures$_6$ (and C_6', C_6'', etc.). Imagine a great variety of functions, including non-monotonic functions.

Contrived as these cases may be, they illustrate the point that different histories of interactiòns (in the formation of the perceptual apparatus of the species) may lead to different ways of perceiving nature.[5] But perhaps someone may doubt it. Creatures $C_1 - C_4$ would pick out the same objects by using respectively words '$W_1 - W_4$' that we pick out when we say 'green'. Shouldn't we demand that '$W_1 - W_4$' all be translated as 'green'? But the issue here is not how we should translate '$W_1 - W_4$' but whether $C_1 - C_4$ in fact undergo thre experiences described above. Has enough been said to suppose that they do so? Are the physiological descriptions given enough? For even supposing that we find some crossings and

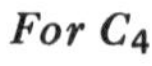

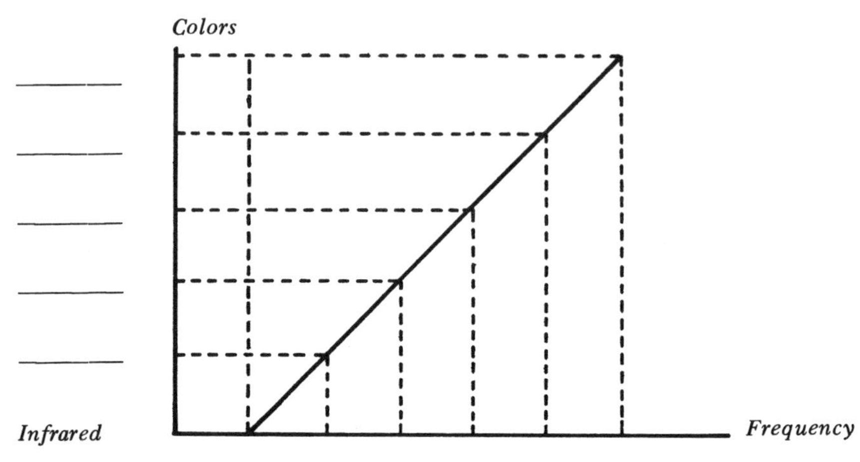

Varieties of correlation between experience and perception

modifications in their neural circuitry as compared to the neural circuitry of humans, since they do pick out the same objects we do under the same circumstances, perhaps we should abandon our original hypothesis, that the particular experiences undergone by an organism depend both on their input and their visual apparatus. This is an unreasonable step. But maybe it is not so unreasonable to believe that what we had considered important neurological features (e.g., an additional stage, a neural 'cross') are not important.

The objection goes as follows. In humans we have noticed certain correlations between features and states of the perceptual apparatus and the experiences (viz. color experiences) that a person has. But if it turns out that Man_2 plus creatures $C_1 - C_4$ have some neurophysiological anomalies (even though in all other biological respects are identical to the average human), and that all of them pick out the same objects per color word that the average human does, why couldn't we suppose instead that we have found counter-instances to the neurophysiological theory? That is, that perhaps the correlations do not hold. And we can stand fast by this position unless there is a way to tell that they do have different experiences.

I believe that this objection is wrong, for it ignores the fact that the complementary color case is just one among many in a wide spectrum. Case 5, for example, would lead to a different range of color word responses (as would all the other non-linear cases). Moreover, consider cases which rather closely resemble humans and

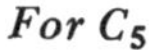

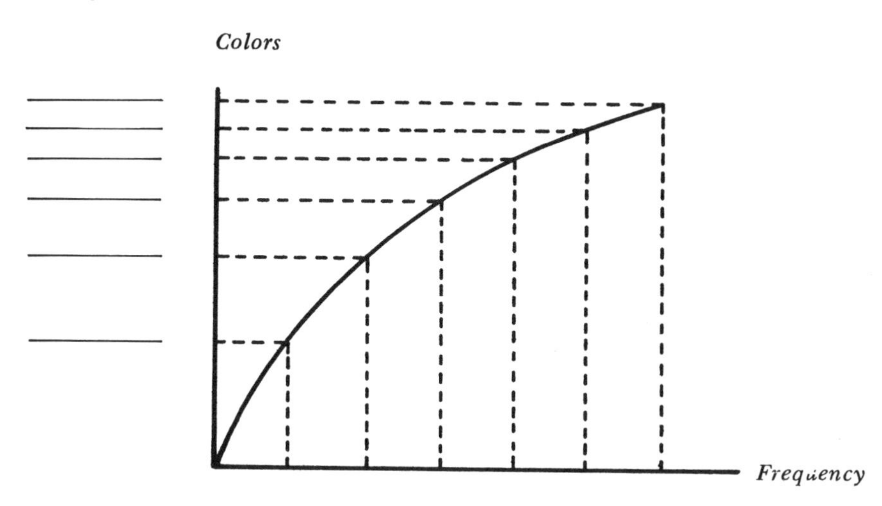

Varieties of correlation between experience and perception

$C_1 - C_4$'s but which do not offer a perfect match (the response curves would not quite overlap). The color identifications made would not be quite like ours (or those of C_1's – C_4's within our range of response). Only a small minority of all cases would offer a perfect match (if some C_1's had *some* of the green and red responses switched we could still tell, for they could not pick out our yellows; only a rare, perfectly balanced switch would give rise to the Wittgensteinian claim that we have no issue).[6] Those few cases would constitute small gaps until filled in, within a general theory of color perception, with instantiations of our color-reversal hypothesis. Such cases, then, should be seen as part of a continuum, they fit well in it. In such a context we can see that there is an *issue*, that to speak of different experiences becomes philosophically suspect only if we take too narrow a view of the matter.

We need not limit ourselves to the availability of the overwhelming theoretical weight in order to crush the objections. For we could test the hypothesis by very many experiments. I will mention two that I have imagined.

Experiment 1: Imagine that Man_2 has not always had his circuits crossed. He has been 'normal' most of his life. Also imagine that a scientist who holds the neurophysiological theory in question wants to test the particular hypothesis of the crossed neurons and its predicted result. He kidnaps Man_2 and performs the operation on him. Man_2 wakes up, goes outside and looks at the familiar yard. 'My God,' he will say, 'the grass is red.' Man_2 may then check with a spectroscope and obtain the same results he had obtained before the operation. (Of course, if he does not see red where he saw green before he would have evidence against the hypotheses.)[7]

Experiment 2: Let us imagine, for the sake of our account, that we have good means for dividing into stages the visual apparatus of all the creatures talked about (analogously to block-diagramming in electronics). Thus for humans there would be stages $S_1 - S_{10}$, for C_2's $S_1' - S_{10}'$, for C_4's $S_1 - S_{20}$, and so on. Now it turns out that the last stage on $C_1 - C_4$ is just like the last stage in humans. We would know that they are alike because they have like biophysical and biochemical compositions, like organizational structures, like measurable outputs (of, say alpha and beta waves and so on) given all stage-inputs.

It also turns out that the states of the last stage are the only ones (of all the stages) that are perfectly correlated with experiences (which humans would characterize as say, 'green' and the others as respectively '$W_1 - W_4$'). That is, normally all stages would be involved, but we might, with drugs or by electronic stimulation, excite the last stage alone and the individual creature will undergo the experience.

What we want to know, though, is whether the experiences resulting from the functioning complete chain are alike in all cases. So we take the output from the next to last stage of $C_1 - C_4$ and connect it to the input of the human's last stage, one at a time. (A case of seeing through somebody else's eyes.) The human will then be able to see what the others 'normally' see (though perhaps the infrared and ultraviolets would still be outside his range). If C_5's and some ofthe C_6's also have like last stages, we could provide a human with experiences similar to theirs.[8]

Now, the most important consequence of the interactionist view presented in these examples is the following. In the case of the color blind person, and in the case of Man_2, we felt justified in saying that those humans were deficient, deviant, or perhaps malfunctioning. Some form of perceptual realism presumably gave us our justification (particularly assumption (2) of chapter 2). It seems, however, that the other species would have as much of a right to claim that they are the ones who see the way things are really like. That is, that grass is 'really' red, or 'really' wu, as much as it is 'really' green.

Some of the species may be older than ours, or better established, more civilized, even far more numerous, and so on. C_1 's could say that it is us who have our wires crossed. So could C_2 's. And whereas we could point out to C_4 's that they have an extra stage that results in a frequency alteration, they could answer that the frequency alteration occurs in us, for we *lack* an indispensable stage. And so on with exponential and other types of response.

A realist may wish to insist that the way things really are can have only one mental representation, but this claim would involve assumptions that are very unreasonable in the light of the above. He would have to assume that the psychology of an organism is in no way related to its biology; or else that biology can in principle supply only one strategy for the appropriate perception of nature (ours). Without implausible empirical assumptions such as these, the sort of empiricism that I have been discussing cannot make a stand. It is ironic that realism should be found wanting not on conceptual but on *empirical* grounds.

As a result of the remarks made so far and in the rest of the chapter, it should become apparent that an observation language will not be neutral between radically different theories because there can be no 'preferred frame of reference'.[9] This I call the Principle of Relativity of Perception. Toward the end of the chapter I will extend it to cover intelligence and science as well. It must be clear that the principle does *not* say that *all frames* are on a par, but only that several *may* be – thus we cannot assume that our is *the* representation of reality. The cases we have examined indicate how it is possible, and perhaps even likely, for *many* frames to be on a par. But these cases also describe visual frames of reference 'superior' to those of dogs and other creatures with bad eyesight.

What we have done for sight can be done, *mutatis mutandis*, for all the other senses, individually and collectively. Our little story has been a very simplified one, though. We could say that the case has been made for sensation only. But the comparisons apply equally well to perception. The distinction being made here is that of classical psychology between sensory experiences which correspond to preliminary elements (sensations), and those which correspond to a secondary synthesis (perception). Such a distinction is no longer held. As Piaget puts it:

> There is immediate perception as totality and sensations are now merely structure elements and no longer structuring . . . When I perceive a house, I do not at first see the color of a tile, the height of a chimney and the

rest, and finally the house! I immediately see the house as *gestalt* and then analyze it in detail.[10]

Furthermore, the elements of experience that we would want to call sensations depend not only on the stimulus and on the perceptual apparatus *simpliciter* (i.e., color not only depends on the stimulus wave-lengths and intensities, and on the retina configuration and so on), but also on whether the patterns are accepted as representing objects. Within a context of recognition, sensations come to full life. A whole range of other factors is also crucial: 'expectation, or previous knowledge, of the normal colour of the objects is important', as R. L. Gregory points out.[11] Some colors such as brown and silver, for example, require contrast, pattern, and preferably interpretation of areas of light as surfaces of objects before they can be seen.

Not only are expectations highly influential in sensory experiences,[12] but we must also say of perception in general that it is highly idiosyncratic at all basic levels, partly with respect to the species and partly with respect to each individual.[13] We have phenomena of color constancy, size constancy, and form constancy, just to name a few. Our perceptual apparatus tends to impose order so as to enable us to manage our environment. There are certain states of equilibrium that we seek. But what may be convenient for us may be highly inconvenient for other species. Not only may it be advantageous for them to have their constancies at different points, but to have different kinds of constancies, i.e. lacking many or all of those we have and possessing many others we lack.

Our eye mixes colors, such that from a few principal colors the full gamut may be generated. Two colors can be mixed to give a pure third color, that is, one in which the constituents cannot be identified. This need not be so. If our optical system were more similar to our acoustical one, such mixing of colors would not come about. Two sounds cannot be mixed to give a different pure sound. The particular perceptions of an organism (of the species) would be the result of an interaction between the perceptual apparatus of that organism and the environment[14] (or better still, though we are anticipating here a bit, of a history of interactions).

What are we to make to the many possible conflicting claims as to the way things really are? As I see it, we now have the picture of many organisms interacting with the universe. The 'facts' depend on those interactions, are their results. And since there can be many kinds of interactions, even for just *one* feature of the universe, there can be many kinds of facts where we thought there should only be one.

The considerations of this chapter lead us to a Principle of Relativity in Perception somewhat analogous to the Principle of Relativity in Physical Theory. Just as in physical theory, any 'measurement' must be relative to the frame in which it is made, and no frame is outright 'better' than any other possible (i.e. no one has a position of privilege over all others). But unlike Einstein's vindication of Galileo's principle, according to which *all* inertial frames are on a par, the Relativity of Perception only tells us that *many* frames *may* be. This should not seem surprising, for whereas Einstein assumes the homogeneity of space,[15] we are not in a position to assume its equivalent in perception, i.e., the homogeneity of the biological conditions, mainly evolutionary, of which the perceptual frames of reference are the results. The main shortcoming, if it is a shortcoming in this case, is the lack of invariants that would permit transformations from frame to frame.[16] In spite of this disanalogy to physical theory, the sort of relativistic epistemology under consideration is armed with a most powerful defense.

Given the Principle of Relativity, no longer should it be considered intelligent to ask whether we see things (the universe) as they really are. For ultimately the way the universe 'really is', or 'really is not', is relative to an interactional system of reference (e.g. the species *homo sapiens*). Otherwise it is like asking what the 'real' mass, or 'real' velocity, of an object is – that is, independently of any frame of reference. Similarly, we do not violate the principle of non-contradiction in saying that grass is red and wu, as well as green, anymore than we violate it when saying that the velocity of an object is nv in one frame of reference and 5nv in another. And the question 'Do you mean that if there were no organisms interacting with external objects, there would be no external objects?' is somewhat similar to 'Do you mean that if there were no frames of reference, objects would have neither mass nor motion?'[17] Thus we dispose of the traditional objections against relativism.

So much for assumption (2), the absolutist notion about *the* (one) *structure of reality*.[18] But what about the ideal of complete knowledge, assumption (3)? The handy theological myth is no longer useful. Not being an interactionist system of reference (not having a 'body') God would have no way of knowing the way things really are, let alone the uncountable possible ways they could be. A creature that fulfills the naive realist ideal (reality-experience) sounds most implausible now. It is as if we had required of our myth that it be acquainted with the absolute mass and motions of all objects, and now we were to realize that such a myth cannot come about because the requirement cannot be met.

Of course, I have made certain assumptions about the diversity of biological realization and the complexity and idiosyncratic character of a species' perceptual apparatus.[19] I have accepted the general thrust of many accompanying hypotheses that are well corroborated. The methodological aspects of my position thus far can be summarized as follows: I have tested my philosophical position against some parts of our empirical knowledge, which has resulted in my making some plausible assumptions and in accepting some hypotheses. I could have made other assumptions such as the ones which, as I claimed, a realist must make in order to hold his view. But such assumptions are implausible. Perhaps I should not want to insist that it never pays to make implausible assumptions, though surely it is not unfair to ask why they should be made in this instance. In any event, the important point is that the recent separation between philosophy and science collapses. What was seen as the independence of philosophy was merely the failure to realize that philosophical positions presuppose certain views of the world. I have not established that all philosophical positions do so, but then it is not necessary to do as much in order to show that the separation thesis was incorrect. This is not to say that there are no differences between philosophy and science, but rather that a very crucial relationship exists between them. And it is difficult to overestimate the significance of this result to methodological concerns in philosophy, and probably in science as well.

The argument against absolutism does not apply only to the realism of classical empiricism and of any other view of science that depends on the uniqueness or the neutrality of the observation language. It also applies to a hypothetical realism extended to observation theories. Thus a Popperian recipe for the continuous overthrowing of observation languages in favor of other observation languages that come increasingly closer to the structure of the world ultimately fails. It fails because such a schema can account for only one line of theories of observation, all of which will presumably be at least partially dependent on the results of one particular interactionist system (or historical line of interactionist systems). And my schema suggests, on the other hand, the possibility of many different such lines, all just as 'good' in their own way.[20] (Insofar as Popper's view is restricted to some 'higher', or more traditional, notion of 'theory', we must wait a few pages. It is not clear that such restriction is warranted, however.)

As soon as we have an interactionist system, certain kinds of interactions (with certain resulting experiences) can come about. Thus from the start we have a 'bias' built in, a 'theory' about how to perceive the universe.[21] As a result, any observation language, inasmuch as it partially depends on perception, which from the empiricist's point of view must be the case, also has a bias built in from the start: it cannot be neutral. This theory of ours, nonetheless, need not be a rigid one (it may be affected by learning to some degree), and its instantiations may vary with exposure to particular environments (social, intellectual, etc.; indeed we should more properly speak of a particular variety of 'theories'). This diminishes the neutrality of the observation language even further.[22]

Expectations, as I have already said, play an important role in perception. Several of them are probably invariant to many species. Some of the optical illusions we are subject to also occur in several species of fish and birds. That is not surprising because those optical illusions are caused by the application of certain forms of perceptual constancies beyond their normal range, and the other organisms in question also possess similar constancy mechanisms. But other such expectations seem to be culturally dependent, they are the result of learning and/or exposure to certain environments. The Zulus, for example, experience some illusions to a small extent, and are hardly affected at all by many other illusion figures.[23] The response can confidently be ascribed to their having grown up in a non-perspective world, with curves instead of straight edges.[24]

Given Kuhn and Feyerabend's view of science, theories are as spectacles through which scientists look at the universe. 'Science is the cooperative perception of the universe', claims R. L. Gregory.[25] Thus scientists approach their work with certain expectations. We can see, then, how far this line of thought may go against the separation of theory and observation. This topic is very interesting in itself and leads to many important issues, but I will not pursue it further at this time. My main interest was to provide a general argument against a certain epistemological model. If it can also serve as a background against which to evaluate particular arguments, so much the better.[26]

My argument does not stop at perception, however. Since I think that, in general, all forms of cognition are the result of interaction, I also think that the Principle of Relativity, a consequence of interactionism, applies to the basic intellectual structures of man as well, whatever they might be. If logic is one of them, we will have a relativity of logical structures. A problem will arise immediately about (1) whether a particular logic, say two-value logic, is just the result of a response to a limited environment (that is, the species has a diverse logical arsenal that it could employ in a diversity of environments);[27] and (2) whether in case it can be established that the logical arsenal has been exhausted by a particular logic, the logical structure of the species may nevertheless change through evolution. This topic will be treated in more detail in subsequent chapters.

I think it is fair to say that throughout the history of philosophy perception and intelligence have been considered very different from each other. So even if the Principle of Relativity applied to observational languages, insofar as they are partially dependent on perception, some work must be done in order to extend the principle to the intellectual realm as well. This I propose to do in two ways. *First*, I will mention some studies on the relationship between 'lower' structures such as perception and 'higher' ones such as intelligence made by scientists like Jean Piaget and Konrad Lorenz. These two thinkers have claimed that their views are of consequence to epistemology. The main consequence, it seems to me, is that they lend support to an interactionist view of intelligence. I will not attempt to provide a summary of their views, interesting as they are, but will suggest how contemporary research on the matter favors my approach. *Second*, I will again offer a general argument to establish both interactionism and relativity, quite independently of any specific psychological theory of intelligence.

According to Piaget, intelligence is 'the form of equilibrium towards which all the structures arising out of perception, habit and elementary sensori-motor mechanisms tend'.[28] Intelligence is not of a kind with other structures, rather it is differentiated by its much wider scope of application in time and space.

> [Intelligence] is the most highly developed form of mental adaptation, that is to say, the indispensable instrument for interaction between the subject and the universe when the scope of this interaction goes beyond immediate and momentary contacts to achieve far-reaching and stable relations.[29]

We may think of intelligence as a goal, but it is hard to say where it begins: 'its origins are indistinguishable from those of sensori-motor adaptation in general or even from those of biological adaptation itself'.[30] Sharp separations do not work; in the end they turn out to be crude, because, for example, intelligence already appears at the level of perception:

> . . . perception itself does not consist in a mere recording of sensorial data but includes an active organization in which decisions and preinferences intervene and which is due to the influence on perception as such of this schematism of actions or of operations.[31]

Given all of this, and given that intelligence engenders science itself Piaget is quite correct in thinking that 'it is therefore natural that the psychological theories of intelligence should come to be placed among biological theories of adaptation and theories of knowing in general'.[32]

According to Konrad Lorenz human cognition is the result of the interaction of two systems. What we know about reality is the result of adaptation, and central to adaptation is the gathering of information.

> . . . all cognitive functions with which we are endowed, indubitably are, like all other adaptive life processes, the function of organic systems evolved in age-long interaction between the organism and its environment.[33]

The forms and categories of conceptualization are also functions of central nervous organization and as such they can be phylogenetically traced back to other nervous structures such as those of perception.[34]

> The . . . methods of comparative phylogenetic research . . . show quite unmistakably how smooth a transition there is between the mechanisms of spatial orientation and perception on the one hand and a priori forms of thought and conceptualization on the other. Despite the enormous differences between these lower and higher cognitive processes in terms of complexity and level of integration, they all quite characteristically fit Kant's definition of the a priori: they are all determined prior to any individual experience, and must be so for experience to be at all possible.[35]

From the point of view of both phylogeny and ontogeny intellectual structures can be best accounted for within the context of interactionism. But interactionism as in the case of perception, leads to relativity. It is not hard to see why one should think of an organism's intellectual structures as being the result of interactions that have a long history. After all, it is no secret that differences in intelligence between diverse species are due to differences in their respective central nervous systems. The brain, for example, is the result of a long, adaptive process. Interactionism can be rejected, of course, but on pain of making some gross and implausible assumptions. But relativity? Let me make the case for it.

I would like to conduct the following thought experiment. Consider Figures A-1 and B-1. Suppose that the space enclosed within

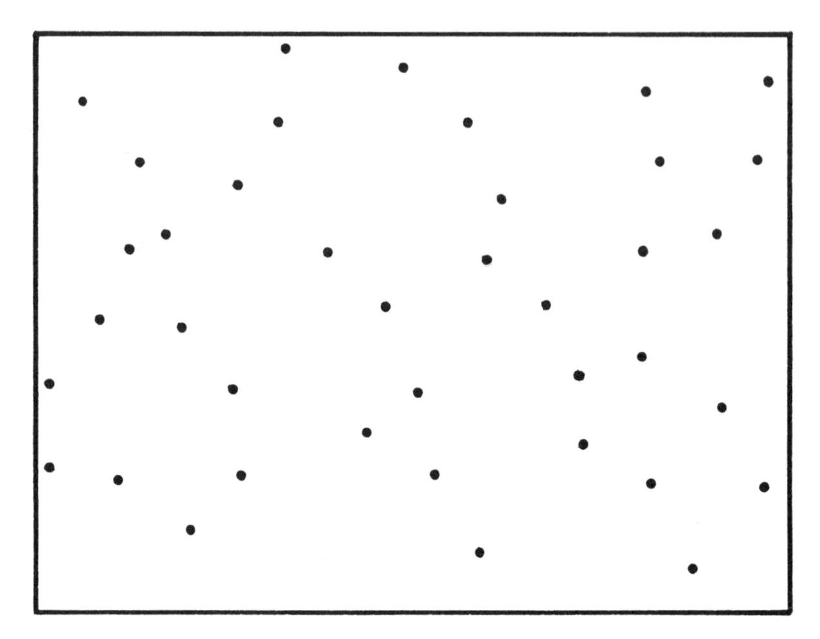

Figure A-1 The A's will perceive the universe as being dotted

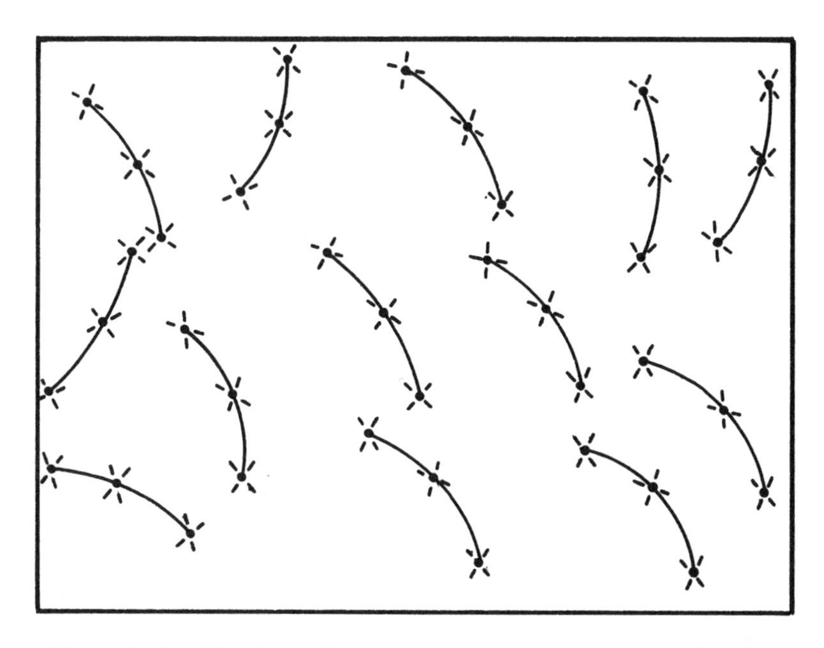

Figure B-1 The B's will perceive the universe as a collection of curved edges (each x marks the place of a dot)

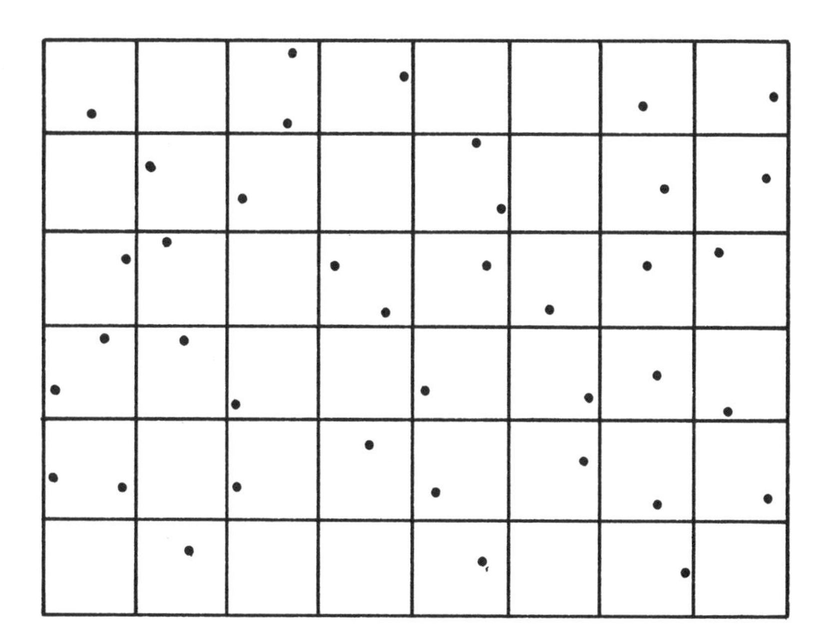

Figure A-2 The rectangular theory of the universe

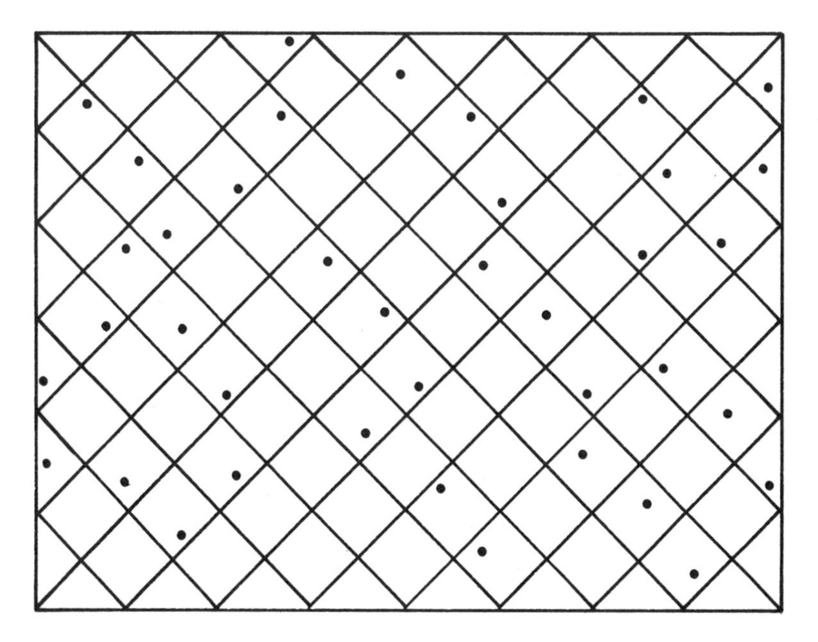

Figure A-3 The rhomboid theory of the universe

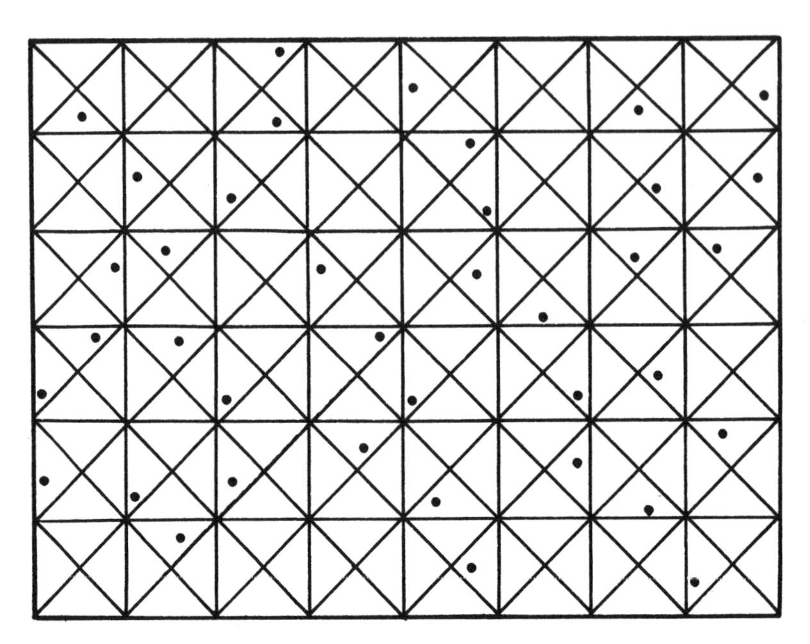

Figure A-4 The A's' compromise: the triangular theory of the universe

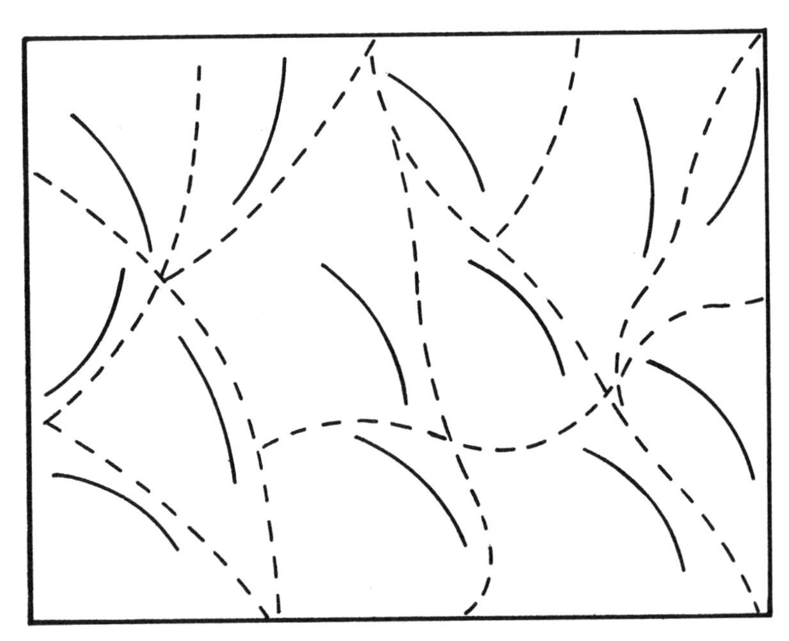

Figure B-2 A physical theory 'natural' to the B's

each rectangle is the universe. Species A perceives the universe as dots. Species B perceives the universe as curved edges.[36] Both species are intelligent. Now, scientific members of species A in an attempt to deal with the universe, viz. with the dots, construct scientific theories. The theories enable them to parcel the dots in such a way that they become manageable. The results are two competing physical views, i.e. the rectangular view of the universe (Figure A-2) and the rhomboid view (which is triangular at the border – an *ad hoc* move according to its opponents, Figure A-3. We could suppose that throughout the history of species A a long struggle takes place between rectangular and rhomboid theories of matter (think of the dispute between wave and particle theories throughout the history of our own species). A compromise is so successful that the triangular theory (Figure A-4) comes to be regarded as corresponding to the ultimate structure of reality (perhaps they might reach the stage of depth which Popper suspects is unattainable for us).

But consider now the situation of the B's. Their perceptual apparatus is just as 'good' as that of the A's, given the considerations about relativity of perception discussed earlier. But taking such perceptual structures as a starting point would make it most unlikely that their intellectual structures, and ultimately their science, develop in the same direction as, let alone be identical with, those of the A's.[37] We can see that neither rectangular nor rhomboid, not even the great triangular compromise can be used to manage the universe as perceived by the B's (try to superimpose diagrams A-2, A-3, or A-4 on diagram B-1). Such intellectual constructions might be so foreign to the B's as to remain forever inconceivable. On the other hand, a framework of curves (Figure B-2), irregular from the point of view of the A's and ourselves, order very nicely the edges of the B's. It will be 'natural' for them to think in such ways, though to the A's it would be bewildering and incoherent.

Once again *the* way the universe 'is' or 'is not' is relative to an interactionist frame of reference. The arguments used earlier against preferred frames apply just as well here. A consequence of this position is that certain kinds of questions are ruled out. Those would be the kinds of questions that would presuppose a distinction such as Kant's between *noumena* and *phenomena*. In fact, we should revise our talk of '*the* way things really are'. Metaphysics, insofar as it makes the substance of such talk its crucial concern, collapses into epistemology.

One could think of logic and some physical views as 'corresponding to reality' (in Popper's sense) by making certain implausible empirical assumptions, of course, just as in the case of perception. But then, as Piaget asks, '. . . may we hope for a real explanation of intelligence, or does intelligence constitute a primary irreducible fact, being the mirror of a reality prior to all experience, namely logic?'[38] To assume that it is such a mirror will go against the grain of what we are led to expect given interactionism. Nonetheless that assumption has been made by hypothetical realists from Mach to Lorenz and Popper.[39] Lorenz, for example, thinks that there must be a convergence at the level of mathematics and physics:

> Certain forms of conceptualization and thought are 'necessary' only to the extent that certain natural laws are so all-pervading that any higher organism must enter the world with the ability to take account of them. Almost every higher animal incorporates in the organization of its body and behavior structures which are adapted for such inescapable facts, e.g., that two solid bodies cannot occupy the same area in space, that light passes in approximately straight lines, and that the effect always appears after the cause in time.[40]

Convergence may be an occasional phenomenon, but the universal character Lorenz attributes to it is not very likely for the following reasons.

(1) Let us suppose that there are indeed features of the universe so pervading that all higher organisms must deal with them. Such features would then be dealt with in a *variety* of ways, each the result of the long interaction of the environment with a particular cognitive frame of reference. To say that there must be only *one* such way is again to grasp for a preferred frame (and to make all the concomitant assumptions). Even if there are such universal features, the resulting intellectual adaptations (the logics, etc.) may still be quite different.

(2) It may be thought that since we cannot conceive alternative conceptual schemes they must be out of the question. But this failure may be purely a reflection of constitutional inability or lack of imagination. Indeed such failure should be expected even in many instances in which alternative conceptions would exist. In my example the A's could not conceive of the B's' forms of reasoning and vice versa.[41]

(3) As will be seen in the following chapter, not all features of the universe will be relevant to all organisms. At the outset we could conceive of such disparity in the perceptual systems that some species can hardly be similarly affected. We must not forget that the resulting systems are the product of a long evolutionary process, and that the historical lines may have gone in increasingly different

directions, perhaps even transforming the environment as they themselves changed.[42]

Incidentally, it is this situation that makes it difficult to provide invariants. Biological systems are much more complex than the simple physical systems that can be represented in the discussion of Einstein's Theory of Relativity. There is a far greater number of variables to consider, for one thing. Sense can be made of my Principle of Relativity by taking into account certain biological considerations.[43] The problem is, however, that biology involves us with evolution, and evolution is a historical discipline. It helps us to understand the present in terms of the past, but it does not predict much. The search for invariants may be very fruitful, nevertheless, at least in posing some interesting problems.[44]

The denial of interactionism involves some very implausible assumptions. And, as we have seen, interactivism leads to the relativity of perception and intelligence. Since scientific knowledge is a product of intelligence, we are also confronted with the relativity of science. But perhaps a critic would like to ask a burning question: this view is stated within a conceptual frame of reference, is it not therefore relative? Furthermore, in qualifying the realism I attack as absolutist it may be thought that perhaps my view is compatible with some other form of realism. After all, Einstein was a physicist even though he went from an absolutist to a relativistic position.[45]

Let us consider the suggestion. Interactions seem to be the basic elements of my view. What would be the status of a theory of such interactions, i.e. of a theory that provided the necessary invariants? Such a theory would have mechanisms for effecting transformations between different cognitive frames of reference, or for pointing out why certain transformations cannot be carried out (in cases of overall structural differences between the intellectual genotypes of the several species in question, which would require the invariants, if any, to appear at more elementary cognitive levels).

In spite of my misgivings about invariants, nothing in my position forbids a theory of interactions. Would not such a theory, then, such a Big Picture, present us with a one-world-one-view situation? And if that is the case, are we not facing a form of realism? Why should we not call it so? Perhaps we are, perhaps we should. I need not be unhappy with an affirmative answer.

Nevertheless realism would have been transformed. I still would have made the point that it is a mistake to think about *the* physics, *the* chemistry, etc. What we would have is a theory of cognitive interactions, *with a lot of give and take*, that will allow for many equally 'good' (epistemologically), and radically different, physics, chemistries, etc. The invariants of such a Big Picture would be *cognitive invariants* (cosmological views, or quantum physics, would not do, for example). Even so, is not the Big Picture an *isomorphic* picture of reality after all? Does it not *correspond* to reality? Maybe. But this picture is not *about* what Popper *et al.* had in mind. (It may have to include the range of cosmologies available to creatures capable of certain logics, say, and so on, but it will not claim, 'this is *the* cosmology, etc.'. Absolutist realism went like this: *the* physics is one picture, *the* molecular biology is another . . . all together these give us a composite – *the* picture of reality.)

We must pause to remember, furthermore, that these allowances presuppose the possibility of a theory of interactions. But the interactions in question would be the result of long phylogenetic adventures. Thus our Big Picture will be an historical, not a predictive one (as in neo-Darwinism). Can we then provide the required invariants? Suppose we can do as well here as in genetics; still how would this Big Picture *correspond* to reality? There are all kinds of problems here.

Even if there were none, I would not be forced to stray from my kind of relativism. To perceive is to perceive in *some* way, to think is to think in *some* way, to conceive a Big Picture is to conceive it in *some* way. That is, the Big Picture itself will arise within a particular cognitive frame of reference. But then there could be alternative accounts of interactions that are just as 'good' epistemologically. Not even at this level would we have preferred frames. Lo and behold: we end up with a plenitude of Big Pictures!

References

1. This is actually very rare. See R. L. Gregory, *Eye and Brain*, McGraw-Hill (1966), p. 126.

2. This case would be just one among many of green-red reversals. A perfect 'match' is required to give us yellow.

3. There are many redundant and overlapping systems in the brain, which leads to the fixing of different neural paths for perception in each individual. The changes suggested by case 1 below would require more than simply 'rewiring' neural circuits; thus if the warning about oversimplification is not taken seriously, the case could be very misleading. The complexity of actual cases can be staggering. The classic on-off model of nerve impulse is just one of many ways for the transmission of messages among the cells of the nervous system. Even hormones may be used to signal and amplify near and remote processes. For an analysis of these mechanisms see Schmitt, F. O., Schneider, D. M., and

Crothers, D. M., eds.: *Functional Linkage in Biomolecular Systems*, Raven Press, New York (1975).

4. N – 700. K is a constant. This is obviously an oversimplified account, it suggests a far more linear function than that in existence: there is no equivalent to K. Hue discrimination increases sharply around yellow.

5. This point is in accordance with the general direction of the neurosciences. In the words of the neuroscientist Robert B. Livingston, 'The behavioral selection pressures that were applied to our ancestors and the talents that emerged in them by virtue of the chance changes in their DNA have conclusively shaped our nervous system, its limitations and its potentialities, including all means by which we perceive, discriminate, and comport ourselves', *Sensory Processing, Perception and Behavior*, Raven Press, New Yor (1978).

6. See note 1.

7. This case may be compared to G. M. Stratton's experiments with inverted lenses. Subjects were at first disoriented but finally became used to their new way of perceiving the world. Fiction has it that their perceptual apparatus inverted the images back to their original position. The comparison is of importance here because it may be thought that, through color constancy probably, Man$_2$ would go on seeing green. As it turned out, however, the images never reverted to normal, the subjects simply ceased noticing their oddity. And, at any rate, the process of adaptation takes several days. For an interesting account of Stratton's experiments, see R. L. Gregory, *op. cit.*, p. 204.

8. This simplified account necessarily ignores the role of expectations and prior knowledge, which will be discussed below.

9. If our perceptual responses were species specific, then our 'perceptual frame of reference' would be constituted by the biological apparatus evolution sifted for human perception. But we will see later that the story is much more complicated; thus we must think of idealizations of systems with which the environment is 'measured'.

10. J. Piaget, *Psychology and Epistemology*, Viking Press (1971), p. 65.

11. R. L. Gregory, *op. cit.*, p. 125.

12. This point applies to pain as well. See Ronald Melzack, *The Puzzle of Pain*, Basic Books (1973).

13. For an account of relevant neurological mechanisms see Szentágothai, J., and Arbib, M. A.: Conceptual models of neural organization. *Neurosciences Research Program Bulletin*, 12:307–510 (1974).

14. Indeed, persons blind from birth who gain sight may find that those with sight live in a 'very distorted world'. See R. B. Livingston, *op. cit.*

15. In the special theory of relativity.

16. Any analogies being drawn, then, are to the first postulate of the theory of relativity. The second postulate (about the constancy of the speed of light) is instructive, nonetheless, because of the way in which Einstein forced the notion of time to take into account the physics of our universe.

17. I hope this result throws significant light on philosophical problems such as the existence of the external world and the related controversy between subjectivism and realism. The way the problem is presented always involves pointing out that we often make mistakes in determining the properties of the universe (e.g., our senses clearly trick us sometimes, thus we can never be entirely justified in trusting them); and since we are never justified in believing that the properties we ascribe to the universe are *the* actual properties of the universe, then we can not be justified in believing in its existence either. That is, it is taken for granted that assumption (1) implies assumption (2), and then failure to justify (2) implies failure to justify (1). But part of what I have done is to show that (1) does not imply (2).

18. Someone may think that my position amounts to a proposal to relativize the notion of reality. That may be all right. But it would not be all right to resist the proposal on the grounds that the relativity, if any, would be purely epistemological (one can see the parallel to physics here). Keep in mind that the issue is the connection between observation languages and reality. Then I will make the following claim: The universe has property *p* if it is possible to correctly perceive it as having *p*. Perhaps we could talk about ideal observers here, but it does not matter. What does matter is that 'correctly perceived' has been relativized. To perceive is to perceive in *some way* (whether the observer is ideal or not); and that *some way* is the result of *interaction* (thus we are led to relativity again).

Suppose my claim is denied (that the universe has property *p* if . . . etc.). What does it mean then to say that the universe has a property *p*? For something to have a property is to be able to interact with other objects in certain ways (to bear certain relations to them). When it comes to perceptions, again, those ways are relative. So the choice is to either relativize the notion of reality, or to give up the realist position about observation languages altogether.

The same argument applies *mutatis mutandis* to the relativity of intelligence and science, which we will discuss later.

19. Adhering more closely to the progress of the neurosciences leads to rather similar conclusions regarding the appropriateness of relativism. See R. B. Livingston, *op. cit.*, p. 11, who also says in his preface that a 'lens-free system of viewing the world does not exist and has never existed'.

20. The word 'line' is not completely satisfactory here. To philosophers perhaps one should speak of disjunctive classes.

21. Again this view is in agreement with the results of neuroscience. 'Altogether, it is evident that the central nervous system can exercise many discriminative controls affecting sensory input. Emphasis has been placed here on the controls affecting receptors because such an early influence would absolutely preclude unmodulated sensory data from reaching perceptual centers. This is a radical piece of information, for it implies the possibility of initial bias affecting data destined for perception at the very beginning of the generation of sensory signals. It thereby hangs a question mark in front of all discussions dealing with epistemology.' *Sensory Processing, Perception, and Behavior*, Raven Press, New York (1978). R. B. Livingston, *op. cit.*, p. 47.

22. According to V. B. Mountcastle, 'Sensations are set by the encoding functions of sensory nerve endings, and by the integrating neural mechanics of the central nervous system. Afferent nerve fibers are not high-fidelity recorders, for they accentuate certain stimulus features, neglect others. The central neuron is a storyteller with regard to the nerve fibers, and it is never completely trustworthy, allowing distortions of quality and measure, within a strained but isomorphic spatial relation between "outside" and "inside". Sensation is an abstraction, not a replication, of the real world.' The view from within: Pathways to the study of perception. *Johns Hopkins Medical Journal*, 136:109–131 (1975).

23. R. L. Gregory, *op. cit.*, p. 161.

24. There are many other cases. For example, the Dutch tend to find the horizontal bar of a T longer, and the Swiss tend to say the same of the vertical bar, even though the bars are of equal length. But perhaps the most striking demonstration of the role of prior experience in perception can be found in Kilpatrick's distorted rooms experiments. *Explorations in Transactional Psychology*, New York University Press, New York (1961).

25. R. L. Gregory, *op. cit.*, p. 225, and Konrad Lorenz, *op. cit.*, p. 259.

26. The following is an interesting point to consider. Presumably our

interactionist frame of reference may be 'correct' (in the sense permitted earlier). In that case, our observation language would serve the purpose of inductivism, falsificationism, etc., would it not? It would then be neutral for human science after all. Well, the observation language may be 'correct', but not in the sense of corresponding to the facts, which was the crucial sense. At best we can have a relativized notion of correspondence. (And a relativized notion of truth as a result: Truth relative not to the individual, society, or culture, but to the interactionist frame of reference.)

The observation language would be 'correct' in a sense that ties it to certain biological considerations that would make such language very likely to change. (It may change by exposure to a different envornmental range, or by the influence of many factors, as we have already seen, including man-made theories). This plasticity of the observation language in the face of theoretical pressure precludes its use as a neutral standard in cases of theoretical change.

I trust it can be readily seen that if 'correspondence' ('truth') is an absolute notion, then no observation language can 'correspond' (be 'true'), for many observation languages may be just as good. And if we are willing to settle for a relative notion, my position supports the claim that observation languages must be theory-laden (or perhaps just plain theoretical, as Feyerabend argues).

27. Speaking a bit metaphorically, we would wonder whether we have determined a particular 'intellectual phenotype' but not the species' 'intellectual genotype'.

28. J. Piaget, *Psychology of Intelligence*, Littlefield, Adams & Co.

29. *Ibid.*, p. 7.

30. *Ibid.*

31. J. Piaget, *Psychology and Epistemology*, *op. cit.*, p. 87.

32. *Ibid.*, p. 11.

33. K. Lorenz, *op. cit.*, p. 255. See also *Evolution and Modification of Behavior*, Chicago Press (1965).

34. *Ibid.*, p. 304.

35. *Ibid.*, p. 288.

36. They would see everything the way cats see some things.

37. Notice that the intellectual structures do not 'mirror' the perceptual ones, but rather that, given the perceptual apparatus of the creatures in question, the degrees of freedom of the intellectual structures are restricted (a 'partial determination', in philosophical jargon).

38. J. Piaget, *Psychology of Intelligence*, *op. cit.*, p. 17.

39. Popper is not of one mind on this subject (see chapter 6). One should distinguish, as David Paulsen has suggested, between the kind of realist I have been discussing and the *contingent* realist, who claims that it is unlikely for radically different sciences to exist because it is unlikely that the beings needed to produce them exist. For the contingent realist, then, our world is the world, since our world is (*de facto*) the only world there is. Such a position accepts my views on what it is to conceive of the universe but finds nature far more impoverished than I do. The remarks made against convergence apply in this case as well.

40. K. Lorenz, *op. cit.*, p. 290. Lorenz hesitates between this position and one closer to mine. See chapter 6.

41. This topic will come up again in chapter 6 and will be treated in some detail in chapter 7.

42. Some may wish to point to instances of convergence on this planet: Cetaceans and fish, for example, or marsupials and certain mammals. But there is much less to convergence than meets the eye, in the first place. In the second place, in the case of ocean animals, to take up one of the examples, the

action of the environment is direct and pervasive. In the case of sophisticated nervous systems, there is considerable evolutionary slack between the genome and the resulting behavior: not all behavior is selected for or against, there is tremendous plasticity, the neural pathways tend to be long circuiting and so on. In the third place, as we have seen, the structures of intelligence arise out of the structures of perception, among others; so we can imagine that a species that has developed in a completely alien world – one that offers dramatically different perceptual opportunities – may develop quite unusual brain mechanisms. It is those brains that will eventually turn to the contemplation of the universe.

43. Professor Lieber has suggested to me that Galileo was brought to the Principle of Relativity by biological considerations. In that case the view has come full circle.

44. Perhaps the approach will parallel that of genetics.

45. Although in some respects Einstein's view is absolutist.

4

THE PERFORMANCE MODEL OF SCIENTIFIC KNOWLEDGE

The arguments in chapter 3 further one of my main concerns: that of presenting an alternative to absolutism in epistemological pursuits. Without a point of view, without a perceptual or intellectual framework of reference, an epistemological theory cannot be connected to cosmological concerns. Protagoras' maxim, 'man is the measure of all things', has been vindicated. But with a twist. Perhaps we should say, 'man is *a* measure of all things'.

In this chapter I will be concerned with another aspect of the most prevalent epistemological ideal, namely, with the demand for a cumulative growth of scientific knowledge. Against this demand I will suggest that scientific merit lies not in the approximating of such a (mistaken) cumulative ideal but in performance vis-à-vis the environment. This suggestion should come as no surprise, for the relativism of the previous chapter implies it. Surely, there are no 'preferred frames of reference' in the illustration in chapter 3 (cases $C_1 - C_6$) only if we realize, as we do, that no one frame enables the corresponding species to deal with the world in a superior manner. Biology may provide a variety of structures and mechanisms for perceiving and thinking about the universe which will serve the appropriate species as well as ours do us (or better, for we cannot be assured that *homo sapiens* is at the pinnacle). But not all frames are on a par, as I mentioned. And within the same species it is clear that a great number of views are possible, and that some at least appear to make fuller use of the intellectual and physical resources potentially available to the species.

The notion of performance arises in the context of adaptation in particular and evolution in general. It has been shown that science depends on intellectual frames of reference, and that such frames result (to a large extent) from a history of interactions with the environment. But there are no strict guidelines on just how close the relationship between a performance model of scientific knowledge and evolutionary theory ought to be. Of course, there is a need to sort this matter out, and that is one of the main tasks of this and the next two chapters. The other main task is to develop an alternative to the cumulative model of science.

4.1 A sketch of the new model

The demand for cumulative growth would turn science into a fact-collecting enterprise; and the resulting view of knowledge may be fairly compared to a process of filling items in a checklist – thus I will dub it the 'list' model of scientific knowledge.

G. E. Moore, in answering the sceptic's doubts about the external world, produced a *list* of 'things we know'. Knowledge is supposed to be *justified true belief.* And beliefs are expressed by *propositions.* So whereas the modern philosophers would have wanted a list of *truths*, the contemporary ones want a list of *true propositions* (for, it is claimed, a true proposition 'corresponds' to a fact).[1]

A complete cosmological theory would be a list of linguistic expressions (propositions, sentences, statements – as you wish) that corresponds to all the facts. Thus, in a sense, being cosmologically omniscient is equivalent to having a language that fits the universe like a glove. But only in a sense, for all we require is the complete list of true propositions about the universe, and surely a language is more than a list of linguistic expressions.[2] In Wilfrid Sellars' view, for example, the function of science is the 'mapping by language (scientific linguistic framework) over events'.[3]

The idea that to have knowledge is to be in possession of many facts, the more the facts the more the knowledge, is certainly not new. This idea gives rise to the *cumulative* view of scientific knowledge. To make progress is to come up either with straightforward discoveries of new facts, or with theories that will enable us to add to our fact store-houses. The ultimate scientist, the cosmologically omniscient creature, would have produced theories with which he could have (and would have) *checked* all the facts (as in a *checklist*); or, better still, he would have 'directly' been able to 'collect' them all. I do think, like Lakatos, that progress and striking out into new areas are closely connected. What I disagree with is the notion that such connection *must* lead to a cumulative view of science. This particular point will be taken up in chapter 7.

The notion in question is, then, that there are facts out there, somewhere, and the task of science is to map over them (the epistemological ideal is to map over all of them), to collect them, and then to continue adding to those we already have. According to Hilary Putnam scientific knowledge is cumulative, otherwise it would be 'hard to see why it would be of any theoretical . . . interest'.[4] Whether we can ever achieve Total Knowledge is another question, one to which the answer is: highly unlikely, if not impossible, given that we are finite, prone to mistakes, and so on.

The crudest form of the *list* model of science can be found in inductivist approaches to the philosophy of science, particularly in the proposals toward induction by enumeration. True enough, induction by enumeration is too slow, too cumbersome, etc. So more useful inductive logics have been sought in order to do the job faster, more elegantly, and so on. But apart from the pragmatic considerations, induction by enumeration remains at the heart of much of inductivism, insofar as it is thought, as Reichenbach did, that such logic would eventually fulfill the *knowledge-extending* function of science or that it would fulfill it if any method would.[5] In the very wording of what is to be done we can see the *list* model in operation: the more facts we collect, check against our list (enumerate), the closer we get to finding those generalizations (abbreviations of found facts) that will lead to correct expectations (predictions). Only if we have the complete list can we have Total Knowledge.

Other epistemologies are more sophisticated. They would not just hold that there is a bunch of 'facts' out there and that the business of a theory of nature is to be mapped over as many of them as possible. (Whoever has managed to collect more 'facts' has more knowledge. Total Knowledge would be a complete collection of the facts.) They hold instead, that as we change our theories our 'facts' may also change, including observational facts (which for many are the only kind of facts). But in the end the list model of Total Knowledge remains as a guiding ideal. Popper, for example, would equate

Total Knowledge with a theory that has maximum verisimilitude; that is, a theory that 'corresponds to all facts, as it were, and, of course, only to real facts'. This is an ideal, more remote and unattainable than a theory with *some* verisimilitude, he thinks, but also a more fundamental one.[7]

The traditional conception of Total Knowledge provides a *model*, a guiding principle, for more modest epistemological enterprises. Given the *list* conception, the goal may be unrealizable because of the magnitude of the enterprise: there may turn out to be just too many facts for the time allotted us. But given a not-so-large list of facts, or infinite time, we should be able, in principle, to attain Total Knowledge.[8]

The sceptic attacked the traditional model and found it wanting. The result was that we could not have 'real', or 'true', empirical knowledge. Such was Hume's challenge. Instead of mending the model or replacing it altogether, philosophers, with a few exceptions, suspected a trick of some sort on the sceptic's part, and even subjected him to abuse. Much of analytic epistemology, for example, has been a series of attempts to charge the sceptic with linguistic treachery. But it was the model, not the sceptic, that was mistaken all along. The sceptic was not entirely successful, however, for his criticisms were not raised in the light of an alternative; thus we went on feeling that there must be something wrong with his questioning.

Let us now take a closer look at my alternative.

To begin my sketch, suppose I have a room where I spend a good many of my hours, where I get on with the process of living. It is my room. And I *know* my room. Why should I claim knowledge of my room? There are many things I can say here:

I can move around in it.
I am at ease in it.
I know what to expect about it (e.g., suppose that when the window is open it is never stable. If it is windy out and I get too close I may get a bump on my head. So I either close the window or avoid coming near it).
In my room I can move my hands without hitting the walls or the door (I am pretty well used to the place). I can estimate how much stuff I have on one side of the room, can go into the closet, etc. At night, with the lights off, when going to the bathroom I walk on the right side, not on the left, because there is a lamp on the left.

After living some time in my room I know how to *get along in it.* That is why I say I know my room. There is nothing uncommon about this use of 'know'. We use it similarly in many cases, for example when we talk of someone knowing another person, e.g., his wife or girlfriend. In a given situation one expects the other person to react in certain ways, or not to react at all, and so on. But, sticking to the room example, not only do I know my room, but I know my room as well as I should want to know it.[9] Does this mean that I know all the *facts* about it (that I have the complete list of facts about my room)? It does not.

The present line of thought is not one more trivial 'linguistic' investigation. I am not merely emphasizing this one ordinary sense of 'know'; for then some analytic epistemologist might say, 'Yes, but strictly speaking you do not know your room. There are many facts you do not have. The philosophical problem is whether we can have knowledge in the philosophical sense (i.e., whether the list model can be achieved)'.

The point is, instead, that given these two senses of knowledge, one kind of knowledge is desirable but the other is not. And the desirable one is the one exemplified by the room example. The philosophical, or 'strict' sense may be a sense all right, but little commends our pursuing knowledge in that sense.

Let me explain. There are many facts about my room that I do not know. For example, unbeknownst to me my room is, say, 5092.8645 miles from Rio de Janeiro's City Hall. It may also be a fact about my room that the scent in it is just like the scent of another room in Greece, or in Spain. One could find myriads of facts about anything. But should I have a complete list of such 'facts' before I can justifiably say (and 'strictly' at that) that I know my room? Of course not. For facts of this last sort have one thing in common: they are *irrelevant*. If there must be a list at all it should be a list of all *relevant*, *important* facts. And *relevant* to *me*, the lodger of the room.

Do I have to know that the porosity of the walls in one corner is more than in the other three? Or that a gap between chunks of paint is not constant throughout? Of course not. All of these are irrelevant to me. But the first could be important to an ant or a spider. And the second could be of great relevance to bacteria.

It has already been said that if we fail to restrict the facts to those which are (thought to be) relevant, the process of empirical knowledge can not even begin. My present claim is that the Baconian demand to collect all the facts need not be met for the *completion* of that process either. And with the question of relevance and importance come goals, purposes, needs, and so on, of a cognitive subject. That is, a subject's epistemology (here we should also speak of an entire species' epistemology) should be tailored around the reasonable demands such a subject may make of its science. The demands are reasonable if, were the science to meet them, the subject would then be equipped to *get along* in its universe in a manner loosely akin to that of the room case.

The room example, however, fails to bring out a most crucial ingredient in my epistemological recipe: *flexibility*. That this ingredient is crucial can be seen from a consideration of open-ended cases. Imagine for the sake of argument that a theory is the analog of the equipment that a man takes when exploring the wilderness. That equipment cannot be so chosen that there is a special tool for every possible contingency, since many circumstances cannot be foreseen and even then there would be far too many. Instead of increasing his chances for comfort and survival, the sheer weight of the equipment would overwhelm the explorer. What he needs is something that he can use in a variety of cases, or that would enable him to construct the tools (new equipment) that changing circumstances may demand).

Similarly, the test of whether a theory (be it about the room, the universe, or another person) constitutes knowledge cannot be met by a list (or a collection) of facts which may or may not be borne out by the subsequent events. But rather, such a test rests (at least partly) on the theory's being a tool which allows the subject to deal with novel situations. A theory must then be a sort of 'core' with which the subject may *generate*, with reasonable ease, an extension of the body of knowledge such that a new situation may be appropriately met.

It should be apparent now that I want to propose a conception of Total Knowledge which keeps within the bounds of reasonable epistemological demands (as illustrated by the room and explorer examples). Epistemologists should expect no more and no less of our attempts to provide an ideal cosmological theory. For if we were to achieve such a conception we would have all the knowledge we should want to have, even though we would not, and could not, have all the facts.

We will see later whether Total Knowledge as now proposed is a poor relation of the traditional version, viz. of a complete, total collection, or list of facts, as it may seem.

We have noticed that a very economical theory (insofar the

number of facts is concerned) might do, as long as it has the capacity to generate extensions of its own body which will allow it to adapt to changing circumstances. This is of particular significance, for it draws a contrast between mere knowing and understanding. The cosmological knowledge we want is a species of understanding. We have Total Knowledge when we have complete comprehension of the universe. An interesting analog of this requirement can be observed in the field of artificial intelligence. It seems that for us to feel justified in ascribing knowledge to an information processing device, the device must demonstrate knowledge of a sort that would have something like our new conception of Total Knowledge for an ideal. That is, the programs, the theories with which we equip it, must deal with the environment in flexible and innovative ways (understanding) instead of by exhaustion of possibilities (going over a checklist, or completing a list). Michael Scriven has analyzed the matter in detail and what he has noted can be extended to scientific knowledge in general.

Scriven studies what he considers special cases of the concept of understanding, such as understanding events, processes, or phenomena (frequency modulation or an error in reasoning), understanding a theory (quantum mechanics), understanding a natural language, understanding an experience (divorce or childbirth, car engines). Several contrasts help delineate the concept: understanding versus ignorance, understanding versus 'mere knowing' (this one is of particular importance, as can be appreciated from my previous remarks), understanding versus misunderstanding, and understanding versus believing (or feeling) that one understands. Most analysis of scientific understanding has centered on the first kind of case. Scriven correctly points out that even here different kinds of understanding are involved. Understanding a particular event, for example, may amount to knowing what its cause is, or knowing what kind of event it is. But understanding phenomena such as frequency modulation or weak interactions involves being able to place them in the framework of scientific knowledge (saying what their properties and causes are, knowing how to recognize them in all their manifestations, etc.). He finds that he can apply here criteria developed in his analysis of understanding theories and the like:

> We ask the subject questions about it, questions of a particular kind. They must not merely request recovery of information that has been explicitly presented (that would test 'mere' knowledge, which must be excluded by contrast 2). They must instead test the capacity to answer 'new' questions.[10]

Scriven suggests his three criteria with the qualification that they are not always appropriate:

> The *alpha* criterion involves at least the following capacities which are not essentially distinct:
>
> (a) the capacity to make and recognize (what are currently held to be) obvious logical inferences from and translations of stored data;
> (b) the capacity to extrapolate or interpolate any general principles learned;
> (c) the capacity to apply the theory to new problem cases to which it is relevant.
>
> The *beta* criterion, a related test of comprehension, . . . seeks signs of organization of the material . . . especially any sign of novel organization; we sometimes find a student's 'grasp' of the theory most impressively demonstrated by a novel axiomatizing or a new identification of fundamental concepts. The stress here, then, is on the capacity to produce novel output (where 'novel' is subject relative).
>
> The *gamma* criterion provides a good perspective or appraisal or evaluation of the theory, its components and its achievements, i.e., (roughly) its relations to other theories and the evidence.[11]

Armed with his analysis of understanding, Scriven tackles the question of comprehension criteria for information processing devices: ' . . . comprehension is nothing more than a dispositional capacity side-effect of an effective information storage and retireval system in a complex, open-ended environment'.[12] An environment is open-ended 'if its content increases with time or closer examination';[13] it is complex 'if its features require more than a *few*, *simple*, *predictable* regularities to describe it completely'.[14] A system for storage and retrieval of information is efficient if it is capable of handling more information, by orders of magnitude, that can be wholly stored (or retrieved) in a usefully accessible way (Scriven defines 'usefully' by reference to the maximal reaction time consistent with the goals of the agent, e.g. survival in a fight with a predator; cost and other considerations are also of importance, of course).

This analysis of comprehension, or understanding, finishes the introductory sketch of my position. The emphasis is on performance, or perhaps on potential performance. A theory provides knowledge insofar as it enables the species that holds it to 'get along' in its universe. Knowledge is not cool and detached, but rather the crucial component in a dynamic interaction with the environment.

4.2 The model

Now, can Scriven's analysis be extrapolated to a view of scientific theories? Scriven provides an account of understanding, applied to information processing devices, whose main global structural features seem isomorphic to those of understanding in science (in my sketch, that is), plus an account of the fine structure (his alpha, beta, and gamma criteria). The isomorphism of the overall structure can be seen from (1) the emphasis placed on being able to deal with the environment (with a small fact base, by the use of redundancies, models, etc.) instead of unrestricted fact collecting (mere knowing), and (2) the flexibility of response (we can see here a parallel to my development of the room and explorer stories).

The isomorphism of the overall structures provides the motivation for attempting a rendering of the fine structure in a similar vein. It must be kept in mind that our concern is the structure of scientific understanding, and thus the criteria that follow are not restricted to theories, even if my wording makes it appear so. Such criteria apply not only to theories in an extended sense (which accommodates Lakatos' research programs) but also hypotheses, observation 'theories' (or languages), and the like. Included is, then, anything that may be considered a candidate to provide scientific understanding. Such a candidate is successful when any of the following criteria (roughly based on Scriven's) is fulfilled:

Alpha': The theory enables a society, presumably as represented by its scientists, to deal (directly) with the universe (or rather with a portion of experience). This involves predicting, retrodicting, etc.

Beta': The theory structures experience, or developes the theoretical or experimental tools required for the structuring, such that the experience can be better placed in the 'causal' network provided by the general theory. This comprises the necessary mathematical, theoretical, or experimental work to fit the general theory to the world, and amounts to what Kuhn has called 'the articulation of a paradigm'.[15]

Gamma': The theory connects experience (the environmental input) to the rest of scientific knowledge (bringing it into the causal network perhaps for the first time). The fulfillment of this criterion indirectly allows the society to deal with the environment. Darwin's theory of evolution might be an appropriate example.

The fulfillment for the first time of the *alpha'* and *gamma'* criteria amounts to the creation of a mature science in that field (Kuhn's paradigm, Lakatos' research programmes).[16] Subsequent fulfillments move the field strongly in new directions. The role of the *gamma'* criterion has not generally been recognized, and thus realizing its importance is one of the good results that come from extending Scriven's analysis. Even though these two criteria give us a vision that we did not have before, the bulk of the scientific work of almost any one period will be centered around the *beta'* criterion, which covers theoretical and experimental articulation.[17] This is not surprising, for the fulfillment of the *beta'* criterion adds reach and grasp to what might otherwise be only vision.[18]

The fine structure of the model – as given by the three criteria – constitutes a rough account of the several direct and indirect ways in which nature may be related to our conceptions of it.[19] This is how, I might say, science connects to the world. But such connection must be understood in a context of evaluation in terms of performance. The notion of performance involved, moreover, must be compatible with the biological basis of the relativism defended in the previous chapter. The notion required, then, is (informally) of the sort suggested earlier: that a theory is 'better' than another if it allows us to 'get along' better in the universe. By this I mean that the theory brings the kind of improvement that generally comes about in the following three ways:

(a) dealing with greater ease with our environment (our 'niche');
(b) increasing the number and diversity of environments that we can deal with (enlarging the 'niche');
(c) coping with a continuously changing environment (which puts a premium on flexibility or response). This last point has been stressed by Scriven and lately by Popper.[20]

Such is my account of scientific knowledge as a species of understanding, and of understanding in terms of a *performance* model.[21] When we attempt to fulfill the *alpha'*, *beta'*, or *gamma'* criteria, we are engaging in science. When the result of fulfilling those criteria is any of points (a) to (c), our science is successful, it constitutes progress. But once again this is 'progress' only in the sense of a better performance vis-à-vis the environment, and not in the sense of approaching any Platonic or even Lamarckian ideal.

This account has so far covered conditions whose fulfillment is required in order to speak properly of 'getting along', of 'progress' in my sense. But such conditions in turn set others about the structural (or perhaps social) properties of science as a discipline:

(1) a mechanism for the generation of alternatives;
(2) a mechanism for the (at least temporary) protection of such alternatives;
(3) a selection mechanism.[22]

(1) The fulfillment of conditions (b) and (c), about increasing the variety of environments and coping with a continuously changing environment, would not be very likely without a mechanism for

generating alternatives. The reasons are as follows. The received view, at least in the history of science, will deal (or at least appear to deal) better with the environment than its competitors – surely it must be regularly perceived to have some advantages over them. But even when this perception is accurate, when point (a) is indeed fulfilled, such a received view is most often designed and perfected to deal with that *particular* environment. Thus it need not be suitable, in its accepted form, to extend the work of a scientific tradition to a larger environmental range, nor to cope with a changing environment. And even in the cases in which it apparently enables the species to get along in that one environment, it is not clear such received view has an insight of divine character. Improvements cannot be forthcoming where we deny ourselves at the outset the opportunity of making a fruitful exchange of views, or at least of sharpening what we already have by confronting criticism that would only come to us from a different perspective.[23] Therefore if a scientific discipline does not provide a mechanism for the generation of alternatives, for proliferation in other words, it is unlikely to respond in the flexible manner required of scientific understanding by the performance model.[24]

(2) To satisfy the function envisioned above, the alternatives must be given a chance to develop, to mature, and to recoup (if applicable). There must be an opportunity to apply the *beta'* criterion to them. This is probably what Feyerabend called 'the principle of tenacity' when describing one of Kuhn's main discoveries about science.[25]

(3) A selection mechanism is implicit in the very realization that some views are more suitable to the environment or to the changing environmental circumstances. Sometimes this selection mechanism may favor one view over all others; but other times it may favor several views over the rest of the field, or merely certain features that many views may have in common. What is important to remember about the selection mechanism, however, is that its standards cannot be specified in advance. And here I mean not only theoretical and experimental commitments, or ontological claims, but also methodological standards. After all, our epistemological picture gives us a world with a changing face, with new exigencies, with new opportunities. It is those exigencies, those opportunities that will lead us in determining what questions we can ask most fruitfully of nature and how we can best try to answer them.[26]

These considerations round up my introductory account of the performance model of scientific knowledge. Further qualifications, specifications, and illustrations will appear throughout the rest of the essay. Before proceeding with that task, however, it is advisable to discuss the issue of scientific rationality.

I think the aim of rationality in science is to ensure progress, or at least to lead to progress if progress is possible. And at first sight it may seem that my account explains how science *could* be a rational enterprise. For we have a model of knowledge, the performance model, which places emphasis on 'getting along' in the universe, that is on behavior that turns out to be most appropriate in the long run. In this manner we begin to see the connection between scientific knowledge and wisdom. Science, of course, does not make the actual behavioral choices, but it puts us in a position to do so (this point will be defended in the next chapter). In any event, being able to improve our lot seems to be one of the results of a science practiced according to the performance model. Is not this a rational enterprise, then, if any is? But can my account really succeed when so many others have fallen prey to the anti-rationalist attack of Kuhn and – especially – Feyerabend?

Of the two main whips that Kuhn and Feyerabend use to drive sympathizers of science from the temple of Reason, perhaps the problem of incommensurability has drawn the most academic blood.[27] Incommensurability results from radical change in science (as in Kuhnian revolutions). It occurs, Feyerabend claims, when 'the conditions of concept formation in one theory forbid the formation of basic concepts of the other'.[28] According to one theory the world is so and so. According to a second theory the world is different. The problem is that we cannot even say that the two at least refer to the same objective situation for they cannot make sense together. Nor can we say that one is better than the other, because such comparison requires common standards of evaluation, and standards may also change as we change world views. Discussion of this second point will be deferred a few lines. The first difficulty is that if science does involve radical change, it seems difficult to deny that the world may change 'because of a change in basic theory'.[29] But this sounds patently absurd! How can science be rational if it leads to such consequences? Many philosophers have argued against the possibility of radical change, but my radical epistemology can hardly find refuge in their quarters. And agreement with Feyerabend seems incompatible with the claim of rationality for science.

This dilemma does not affect my position, however. The issue of whether the world changes when we change basic theory does not arise within the relativity of science. To conceive of the world is to conceive of it within a frame of reference. To ask whether the world really changes when we change frames of reference is comparable to asking (in the Special Theory of Relativity) whether the mass really changes when we change frames of reference. It just not a sensible question.

The issue of the changing standards presents a different kind of difficulty. According to the traditional position on this issue, to be rational is to live up to standards of rationality. In the case of science those standards are presumably given by methodological rules (such as 'reject hypotheses that are in conflict with the facts', 'do not make *ad hoc* moves', and so on).[30] Feyerabend has argued that if this position is correct, science cannot be a rational enterprise.[31] The point of adhering to a method is presumably that the method will lead to success (if anything will); and surely science is a greatly successful enterprise, as one can learn from the grand episodes of its history (Galileo, Einstein, and so forth). Unfortunately in many of these exemplary instances of success the great scientists involved not only violated the method but had to if success were to come at all. By historical and epistemological analysis Feyerabend shows that method (as understood by rationalists) may become an obstacle to success, to progress (as understood by rationalists). 'Good' science is opportunistic in matters epistemological. Thus from the rationalist's point of view it looks as if anything goes. This epistemological anarchy, it is thought, would not only empty the temple of Reason but raze it to the ground!

I have not seen any satisfactory refutations of Feyerabend's arguments and historical analysis, so it seems that I must face them. The problems they create for my account may seem particularly acute, not because they are so much at odds with mine but rather because they are so similar in many respects. My account requires that neither proliferation of alternatives nor tenacity be ruled out, that the mechanism for selection depends on changing circumstances and thus cannot be specified in advance, that the standards of scientific rationality which presumably make up that mechanism are then subject to change as well. But how could this be a rational position? If there are no common standards of evaluation one cannot say that scientific change has been for the better.

Nonetheless I had argued as if the performance model of knowledge made science look eminently rational. Could it be that by a subtle rearrangement of the whole one can see the matter of scientific rationality in a different light? Some philosophers of science (e.g., Lakatos) have tried in recent years to incorporate some of the crucial insights of Kuhn and Feyerabend into their scientific methodologies. Feyerabend has argued, however, that those new methodologies may put on the air of rationality but actually offer to the beleaguered rationalists no more consolation than he does (Lakatos is portrayed as a 'fellow anarchist').[32] Does my account offer more?

I think that the picture that Feyerabend presents of science is largely correct (as well as I might, since some of the essential features are also derived from my model). But I do not think that the conclusions he draws from that picture are correct. The key to this whole issue is the assumption made both by Feyerabend and the rationalists: that to be rational is to adhere to certain standards (or method). The question of rationality is thus resolved by looking into the behavior of individual scientists in the context of certain theoretical considerations. (Did so-and-so follow the method? Had he followed the method could he have arrived at the same celebrated views? etc.) But is this a proper assumption to make? Why should the rationality of science be determined at the level of the individual scientist?

Science is a communal enterprise that tries to gain knowledge about the world. According to my model this task amounts to enabling the species to 'get along' in the world. The question of rationality should then be determined by whether science is structured so as to carry out its function. Rationality ought to be considered a structural property of science, just as freedom (when it exists) ought to be considered a structural property of certain societies. Such societies do not cease being free because some, or even many, of their individual citizens are not open-minded, or because they think poorly of their fellows' ideas and life styles. A society is free only when their citizens do not interfere with their fellows' pursuits, no matter what they think of them, either because there is a tendency to live and let live, or because they fear the arm of the law if they so interfere. Freedom is a structural property that some societies have; it functions like an iron railing on the entire society.[33]

Similarly, rationality is a structural property of some intellectual enterprises. It is present, according to the performance model, when the discipline of science is organized so as to channel imagination, ingenuity, and effort into the activities described by the *alpha'*, *beta'*, and *gamma'* criteria in such a way that the result enables the society to 'get along' in the universe (points (a) to (c)). And what satisfies this social conception of scientific rationality if not, according again to the model, the considerations about generation of alternatives, and so on? But these considerations are the same that Feyerabend finds so revealing about the nature of science!

It does not matter, then, that some individual scientists stick to their initial view, no matter what; or that some hold on to the most successful view of their time and look down upon all others; or that some are always searching for alternatives; or that still others rave against the received view, violate method, and perhaps become historical figures thanks to that. What matters, as we can see from the social conception of rationality, is that the scientific enterprise be appropriately organized to face the surprises, to search for the treasures our diverse universe has in store for an intelligent species. This it does when it fits the performance model of knowledge. Of course one cannot hope for a perfect match, but a communal enterprise comes very close to fitting the model when it resembles science as Feyerabend describes it.

References

1. G. E. Moore, in 'A Defence of Common Sense', *Philosophical Papers* Collier (New York, 1966).

2. Total knowledge should be a list of sentences, one per each fact, according to this model. But can there be a sentence per each fact in the world? To ask the question in a slightly different way: Can the set of all facts in the universe be on a one-to-one correlation with the sentences of a given languages? Are these two questions equivalent? If they are we are confronted with a serious difficulty. All the sentences of a language can be effectively enumerated. That is, the set of sentences can be put on a one-to-one correspondence to the set of rational numbers. Can all facts in the universe be so correlated as well? Let us suppose that all physical events can be appropriately described by their three-dimensional coordinates (Xi, Yj, Zk). If it is so then the set of all physical events cannot be put on a one-to-one correspondence with the set of rational numbers. Thus the set of all physical events cannot be put on a one-to-one correlation with the set of sentences of any language.

3. See, for example, Wilfrid Sellars' 'Scientific Realism or Irenic Instrumentalism? ', *Boston Studies in the Philosophy of Science*, Volume 2, p. 203.

4. Hilary Putnam, 'How Not to Talk About Meaning', *Ibid.*, p. 207.

5. Hans Reichenbach, *Experience and Prediction*, The University of Chicago Press (1938), and *The Theory of Probability*, University of California Press (Berkeley, 1949). Whether *all* inductivists should be included is not clear. Nelson Goodman, for example, would be a difficult case.

6. I have heard of another requirement expressed in a rather strange way: in order to know you have to know that you know. But then, how could you know that you know unless you know? And if you know, you know. Thus, if your knowing entails your knowing that you know, the requirement is pointless. If it does not, it is absurd.

7. Karl R. Popper, *Conjectures and Refutations*, Harper & Row (New

York, 1968), p. 234.

8. Some interesting objections may be brought up against the possibility of ever achieving Total Knowledge, given the *list* conception. Suppose we want to be cosmologically omniscient. Thus we should be able to represent every fact in our model of the universe. A human brain would obviously be too small. Could a supercomputer do the job for us? It could not, for insofar as the computer is part of the universe, it could not duplicate the whole of the universe. If it did, then we would have twice as many facts as before, thus we would need another supercomputer (a super-supercomputer) to account for the facts of the new universe (those of the old universe plus those added by the first supercomputer), but then . . . Or we might say that there would always be a residue of facts unaccounted for by the computer, mainly those that constitute its own representations of the facts of the universe. I think this is the sort of argument that Mike Williams had in mind circa 1972. I understand that I. Hayek had a similar argument roughly on the grounds that a brain could not understand itself (it would be unable to represent *all* of its atomic components plus the relations between such components). This is much less impressive than Williams' argument, for human capacity for modeling need not be limited to the employment of just *one* brain. Objections of this kind obviously assume the list conception.

9. I am assuming ordinary purposes in this example. This and what follows in the text indicate an important role for purposes and the like in the determination of understanding.

10. Michael Scriven, 'The Concept of Comprehension: From Semantics to Software', *Language Comprehension and the Acquisition of Knowledge*, Roy O. Freedle and John B. Carroll, eds., V. H. Winston and Sons, Inc. (1972), p. 32.

11. *Ibid.*, p. 32. Italics and layout are my own.

12. *Ibid.*, p. 35.

13. *Ibid.*, p. 36.

14. *Ibid.*

15. Kuhn would not allow for novelty at all these points, however, for the *beta'* criterion would better be described, in his terms, as 'articulating the known', i.e., fitting the paradigm to the world. One may argue, however, for novelty of organization at the very least (as Scriven does).

16. As a fulfillment of the *alpha'* criterion, consider as examples Newton's mechanics, Lavosier's discovery of oxygen, Bohr's atomic theory, or Watson and Crick's discovery of the double helix.

17. As examples of the *beta'* criterion, consider the development of the calculus and other mathematical tools to aid physics in the period between Newton and Einstein, the refinement of purification techniques in nineteenth century chemistry, and the improvement in particle accelerators in this country.

18. As Kuhn points out, there are immense difficulties in developing points of contact between a theory and nature. Even Newton's accomplishments in mechanics (derivation of Kepler's Laws of Planetary Motion, explanation of the moon's deviation from them, a comprehensive account of pendulums and the tides, and even a derivation of Boyle's Law) were small in number compared to the presumptive generality of his laws. There was much work to be done in adapting his work for application in many other fields. For a more detailed account, see Thomas S. Kuhn, *The Structure of Scientific Revolutions*, Chicago University Press, 1970, chapter 3.

19. The criteria are not all that clear-cut: in some cases there might be a combination of two or even three criteria. I think it is clear that Darwin's is an example of *gamma'*, but Einstein's may involve both the *alpha'* and the *gamma'*, for even though he continued a line of thought that was successful in dealing

with nature, he also did it by bringing about a synthesis of disparate approaches in physics.

20. Popper speaks of the worl as an open system, a point that he emphasizes throughout *Objective Knowledge.*

21. In its connection to behavior this account differs from the philosophical and psychological behaviorism so prevalent in this century in the importance it assigns to the mechanisms that produce knowledge. In my account one can see that the relationship between such mechanisms and behavior is a dialectic one. Our brain system is built for action, as R. B. Livingston points out. Thus, '*behavior is not only the goal of sensory processing, it is important in the shaping of sensory processing*'. *Sensory Processing, Perception, and Behavior*, Raven Press, New York (1978), p. 40.

22. Stephen Toulmin has attempted to account for similar points within the context of evolutionary epistomology. See his *Human Understanding*, Princeton University Press (1972). There are some important differences between Toulmin's position and mine, as will be seen in chapter 6.

23. The proliferation of alternatives as a crucial condition for the advancement of science has been advocated in most of Feyerabend's major papers (e.g. 'Explanation, Reduction, and Empiricism' in *Minnesota Studies for the Philosophy of Science*, II, and 'Against Method', *ibid.*, IV. Feyerabend has also hinted at the evolutionary character of epistemology, and has speculated a bit on whether argument itself has mainly an adaptive function, e.g., 'Consolations for the Specialist', in *Criticism and the Growth of Knowledge*, I. Lakatos and A. Musgrave, eds., Cambridge University Press (1970). For his fullest account of the role of proliferation see his *Against Method*, NLB, London (1975).

24. It has been argued that science grew in the Western Civilization and not elsewhere because only the Western Civilization has had a tradition of criticism, that is, a tradition that allows for, and sometimes even encourages, the generation and development of alternatives to the standard view. Such a critical tradition was absent in China, for example. For a nice illustration of this point see Toulmin, *op. cit.*, p. 218.

25. In his 'Consolations for the Specialist', *op. cit.*

26. One would generally expect the selection mechanism to preserve the standard view until an alternative is developed such that it can lead the scientific discipline to progress (in my sense). The evolutionary emphasis is not incompatible with Lakatos' insistence on obtaining *new facts.* Simplistically: an alternative becomes serious when it predicts new facts, new *kinds* of facts (theoretically fruitful). But its acceptance should also require the 'harvesting' of at least some of the *crucial* new facts (empirically fruitful). All the understanding criteria (*alpha', beta', gamma'*) can be applicable in this connection. Lakatos' account is quite limited, however, and also fails to explain the rationale for empirical growth.

27. Accounts of the problem can be found in all the works by Kuhn and Feyerabend mentioned above, particularly chapter 9 of Kuhn's *The Structure of Scientific Revolutions* and chapter 17 of Feyerabend's *Against Method.* See also below.

28. *Science in a Free Society*, London (1978), p. 68.

29. *Ibid.*, p. 70.

30. Of course, such rules are presumably applied in conjunction with more specific rules concerning the standards of experimental rigor, etc. in the appropriate field of enquiry.

31. Particularly in *Against Method.*

32. See chapter 16 of *Against Method.*

33. *Science in a Free Society*, *op. cit.*, p. 30.

5

DEFENSE OF THE MODEL

An appraisal of the performance model of scientific knowledge should not be carried out in a vacuum. One should keep in mind that the traditional model – the 'list' model – suffers from crippling disadvantages by comparison. Two problems are of particular significance. First, the lack of flexibility in the epistemological ideal at its heart; this shortcoming, as we saw in the examination of artificial intelligence, would not merit such model the title of understanding. Second, its connection with the epistemological myth (direct acquaintance, etc.) that was attacked in chapter 3. Nevertheless, the performance model does not win merely by default. Some objections to it come rather naturally, and thus a response is in order.

5.1 Objections

Two general kinds of objection will be discussed in this section. The first one is that by tying understanding to performance we seem obliged to include certain discredited views into the group of those disciplines that have achieved scientific understanding. That is, that my position permits too much. The second is that in some other instances it may instead permit too little. The heart of this second objection is, as we will see, that any connection between scientific merit and survival value is unwarranted.

To bring the first objection home I will consider a contrast made by Stephen Toulmin between the astronomies of ancient Greece and Babylon. Toulmin related the case in order to show that predictive power does not constitute scientific understanding. With a few suitable changes it will serve the present purposes as well. I must warn that the following reading of Babylonian science is open to question, but I must also point out that only if it is read in this manner do we seem to have a counter-example at all. After dealing with the example, I will return to the matter of the historical interpretation of it.

The Babylonians had extremely well worked out 'tables' for the forecasting of planetary motions, moon eclipses, and other like phenomena. Their calculating techniques were of a precision not even approximated by Greek astronomy until the second century B.C. The first Greek astronomer to rival them in predictive exactness was Hipparchus of Rhodes, who was in a position to borrow from their work and probably did so.

Since their astronomy enabled the Babylonians to know incredibly well what to expect of the heavens, it also placed them in a position to take better command of situations related to astronomical phenomena. It seems, then, that their astronomy was superior to the Greeks', since, thanks to it, the Babylonians could 'get along' in the universe better than the Greeks could.

A closer examination of the nature of both astronomies leads to some apparently perplexing questions for someone who holds a position of the sort I advocate. The Greeks were mainly concerned with understanding the nature of heavenly phenomena, and not terribly worried about generating predictions. Their speculations were oftentimes naive but sometimes a bit insightful (e.g. the moon did not have any light of its own, but borrowed it from the sun).

The Babylonians, on the other hand speculated little if at all. As Toulmin puts it:

> . . . they computed the celestial motions in a purely arithmetical way. Like men who prepare tide-tables, or economists working on 'time-series', they analysed each of the celestial motions into a set of independent variables, each changing in a regular, predictable manner. Once this was done, they could calculate the variables separately, and recombine them so as to determine beforehand (or after the event) on which days in a given year the new moon would appear for the first time, and whether at a particular opposition between the Sun and Moon there would be a lunar eclipse.

They succeeded in extending this kind of calculation to the movements of the major planets, and tried with less success to apply it also to earthquakes, plagues of locusts, and other omens. Lunar eclipses proved to occur in a regular and predictable way, but plagues of locusts and earthquakes were intractable. How they explained this fact, we do not know: no theory has been found either about the things they were able to predict, or about those which they were not. Both the successes and failures of their forecasting techniques remained at the time unexplained.[1]

If this account is accepted, it is difficult to believe that the Babylonians had much understanding of the universe. But it was said before that their astronomy enabled them to 'get along', to respond to their environment rather well – better than the Greeks, at any rate. Thus there seems to be a cleavage between understanding and performance.

It will not do, in order to defend my position, to disown early Greek science as too primitive, and thus not a fair test of the evolutionary (or perceptual, or empirical) conception of scientific knowledge. For the rather thoughtless computational techniques Toulmin ascribes to the Babylonians continued, for a long time, to be more accurate than those carried out in accordance with the principles of sciences we still admire. Even Newtonian astronomy proved inferior in that respect until Laplace's *Mécanique Céleste* greatly improved the situation. Newton's theories, however, could explain *why* those calculating procedures worked as well as they did, and made sense of the observed regularities. Such theories provided us with great understanding, and constitute one of the most extraordinary intellectual tools in the history of mankind.

There does seem to be a conflict here, for a good theory, in this sense of 'getting along', would not require understanding; at best, understanding would be the icing on the cake, quite the opposite of my claim that it is the crucial mechanism for 'getting along' in the universe.

I believe the apparent conflict can be solved by restating in some detail the connection between managing in the universe and scientific understanding. By 'getting along' I meant in part the capacity for a flexible response to an environment that may be always changing, to an open-ended environment; that is, the capacity to meet novel situations successfully, to anticipate problems, to deal with them in a reasonably short time, to gain some elbow room, so to speak, since a rigid, slow response may well rule out survival, a minimum test of the capacity for 'getting along'. It should also be clear that the criteria is not one of individual success or failure, for it has been stated that science is a cooperative perception. We are concerned with the capacity of a species to 'get along' as a whole. Thus applications of the criteria expand not only in numbers and complexity but also in time. Only over a very long period – in some cases when all the ships are in – can we evaluate whether its body of science enables a species to 'get along' in the universe.

In this light (and with the help of hindsight) it can be seen that Babylonian astronomy does not present a problematic case. Not just that it is limited in the scientific range and represents only a small time slice in a long history that may take billions of years to complete. But more importantly: it only enabled the Babylonians to operate in a restricted domain, and to respond in a rigid manner. It is crucial to see that it is not a coincidental matter that the successful application of their calculating techniques was so limited. It remained successful only in that restricted field because it could not be generalized. Furthermore, it was not at all promising, for there was no way of successfully extending it to new problems and the like. The Babylonians, in fact, had no inkling about what to do in order to make their technique into a more flexible tool (another requirement for 'getting along' in the universe). They applied it to anything in sight, including plagues of locusts and political events, with no success whatever. The inflexible application of the same technique throughout can hardly be compatible with some of the reasonable demands that may be placed upon a scientific enterprise: that it may (successfully) extend its domain to new environments and that it may be able to cope with a changing environment.

It should be clear, then, that even though Babylonian astronomy had certain practical advantages for a while, it is not a case of 'getting along' in the universe as suggested earlier in the chapter. What it lacked is precisely what was demanded: a capacity for flexible, generalizable responses to a dynamic environment. We can see that scientific understanding provides what is desired: features, patterns, models, networks of connections that can be applied again and again, a saving, a storage, of information that can represent an enormous array of data, thus enabling a quick response when it is crucial for survival, and so on. As Scriven puts it: 'The best way to handle a rich [informational] environment is to understand it'.[2]

Again we can see the parallel to the field of Artificial Intelligence. The search for an intelligent computer could be categorized as the search for a program (or combination of programs) of general application to an environment of great diversity. Many programs of great precision have been developed, but unfortunately they can only be applied to small, restricted domains. They are special purpose programs without any means of successful extension. Even if we equip a single computer with thousands of such special purpose programs it will not be able to 'get along' in our world. (That is, even if we overlook the enormous problems caused by memory capacity and storing and retrieval times, which would besiege a computer of such magnitude.)[3]

It is not clear that Babylonian astronomy was as thoughtless an enterprise as Toulmin makes it out to be. To be sure the theoretical core of Babylonian science has not been found, but that is no reason to conclude that there was not any. I think it can be surmised from reading Neugebauer's *The Exact Sciences in Antiquity*[4] (which seems to be Toulmin's source) that heavenly bodies were considered divine by the Babylonians, and thus took their place, along with many other phenomena, in their complex religion. It may turn out that the theoretical core could be highly mythical, and unpalatable to our modern tastes, but nonetheless a theoretical core (thus perhaps providing Babylonian science with greater flexibility than Toulmin's account allows). As a result, the characterization of Babylonian science in the example may have been very unfair. But if we do provide a fair characterization, then, we would have no counter example. I suspect that we would confront similar situations with other 'counter-examples' we may find in the history of science.[5]

The first objection, as we have seen, was to the effect that my account of scientific knowledge allows for too much. The second is to the effect that it allows for too little. If getting along in the universe is ultimately the telling point in a scientific evaluation, we are tying understanding to survival value. But a scientist, the objection goes, is not concerned with the survival value of the theories when doing science. In fact, it is often impossible to tell what practical applications, if any, a particular hypothesis, theory, or even branch of science will have. What motivates a scientist is the desire to get to the truth, his intellectual curiosity. The same point applies to the acceptance of theories: the scientific community prefers (or should prefer) those theories which best satisfy its intellectual curiosity, i.e., which are nearer *the truth*, independently of their practical applications.

It must be said that a scientist could be motivated by all sorts of things: money, glory, metaphysics, naughtiness, a demanding father, and so on. If we accept the performance model, shouldn't we rule that pure mathematics, for example, is not a science? For after all, what practical applications, what survival value, do highly theoretical mathematics have? The point of the objection is this: a thinker develops set theory or a non-Euclidian geometry, or even some theories in physics, because of many different reasons, and they get accepted or rejected also because of many reasons, among which we sometimes do not find anything related to practical applications of any sort. The idea is, then, that without such practical applications there could be no connection with survival value. As a matter of fact, what practical applications, what survival value, can Darwin's theory of evolution itself have, since it does not even generate predictions?

I think the objection exemplifies too narrow a view. The notion at play here is that survival value is more or less connected with foreseeable application. It is often said, for example, that whereas animals can only take care of immediate and pressing problems, i.e. react to them, we can behave in ways that do not constitute a reaction to any compelling demands of the environment, we are endowed with curiosity (a higher form of which provides much of our scientific motivation), and curiosity liberates us from the drudgery of plain animalhood.[6]

In response one might be tempted to claim, as Popper does, that science has its origin in problem solving. Scientific curiosity would thus be directed by responses to difficulties generated by the environment (initially, at least).[7] But this saving move is not very promising. As Konrad Lorenz has shown, curiosity exists in animals as 'low' in the scale of intelligence as the Norway rat and the raven. These animals carry out active investigations of the environment, and it is quite evident that such investigations are *not* direct responses to environmental needs.

> On observing a raven with a novel object, first conducting exploratory 'security measures' and then trying out one after the other all the instinctive motor patterns concerned in predation, one is at first inclined to think that the bird's entire activity is ultimately to be interpreted as *appetite behavior* for food-uptake. However, it can easily be shown that this is not the case. In the first place, the inquisitive investigation is at once abandoned when the raven is genuinely hungry: it immediate turns to an already familiar food-source. Young ravens exhibit their most intensive phase of curiosity behavior immediately after fledging, i.e., at a time when the youngsters are still fed by their parents. If they become hungry, they follow the parental bird (or human foster-parent) in an insistent manner, and they *only* exhibit interest for unknown objects when they are satiated. Secondly, when the raven is moderately, though still demonstrably, hungry, the appetite for unknown objects prevails over that for the best available food.[8]

The connection between curiosity and survival value can be made on other grounds, nevertheless. As Lorenz remarks:

> If one offers a tit-bit to a young raven actively engaged in investigating an unknown object, the tit-bit is almost always ignored. In human terms, this means: the bird does not *want* to eat, it wants to *know* whether the particular object is 'theoretically' edible.[9]

Animals such as the Norway rat, and ravens (and men) are actually less specialized than other representatives of their respective zoological subdivisions (we may even think of them as 'more primitive'). Lorenz calls them 'specialists in non-specialization' and 'cosmopolitan'.

> The species-preserving function of this appetite for the unknown and this experimentation of all conceivable behavior patterns of the species is easily discernible. The specialist in non-specialization actively constructs its environment for itself, whereas an animal with more extensive special adaptations in morphology and innate behavior is largely born with this knowledge of the environment. In the environment of a 'specialist', such as the great crested grebe, virtually everything which is of biological relevance — the water surface, the prey, the sexual partner, the nest-material, and so on — is determined by highly differentiated innate releasing mechanisms in the species. Learning is predominantly restricted to the localization of stimulus situations which are of significance for the species. There is no capacity in the range of its abilities for self-condition to alter any feature in these 'a priori' inherited and species-specific conditions of the environment.[10]

And then he provides the crucial contrast:

> . . . the investigative non-specialists are always equipped with very few and extremely broad (i.e. character-*restricted*) releasing mechanisms and a relatively small number of innate motor patterns. It is quite characteristic of the latter that, as a direct result of their minor degree of specialization, they have great versatility of application. Through the fact that such animals at first treat everything novel as if it were of the greatest biological importance, they inevitably become acquainted with every small detail of the most extreme and varied ecological niches which can contribute to the preservation of their existence. *Literally all higher animals which have become 'cosmopolitans' are typical non-specialized 'creatures of curiosity'.*[11]

A specialist in non-specialization can adapt himself to a great variety of environments (and presumably to a changing environment). This is the key to the survival value of curiosity. We can see that the satisfaction of the *beta'* and *gamma'* criteria, which deal with the articulation and restructuring of the scientific network, may well provide us with survival at a distance, so to speak — an adaptive situation not unlike that which results from curiosity in animals. The connection between science and survival, via curiosity, need not require immediate response to pressing demands by the environment.

Instead of foreseeable applications, in questionable cases, we should look at the function those activities have in the network of science (what criteria they fulfill, and so on).

If science is an attempt to satisfy intellectual curiosity, then it seems that the queen of our intellect was not born in problem-solving but in play. To continue with Lorenz's analysis:

> The young raven conducting its 'investigations' is not motivated to eat, and in the same way a young Norway rat repeatedly dashing back to the entrance of its retreat from various points within its range is not motivated to flee. This very *independence* of the exploratory learning process from momentary *requirements*, in other words from the *motive of the appetite*, is extremely important. Bally (1945) regards it as the major characteristic of *play* that behaviour patterns really belonging in the area of appetitive behaviour are performed 'in a field released from tension'. As we have seen, the field released from tension – a *sine qua non* for all curiosity behaviour just as for play – is an extremely important common feature of the two kinds of behaviour! [12]

There is considerable difference, however, between the curiosity behavior of animals such as ravens and that of man. And the difference lies in the fact that man's investigative behavior is pursued until the onset of senility. In other animals such investigations are restricted to an early phase in individual development. Curiosity ends when play behavior ends (the relationship between the two was pointed out above). Lorenz makes a nice case for the thesis that the neotonous nature of man is to a great extent responsible for the very possibility of science, though he may go a bit too far:

> All purely material research conducted by a human scientist is pure inquisitive behaviour – appetite behaviour in *free operation*. In this sense, it is *play behaviour*. All scientific knowledge – to which man owes his role as master of the world – arose from playful activities conducted in a free field entirely for their own sake.[13]

Some people may be sceptical about Lorenz's analogies between animal and human behavior, but I think that at least in this case his account should give rise not only to some modesty on our part, but to a realization of the role of curiosity in the process of evolution. If along the way some other myths about the uniqueness of man (e.g. his high specialization and his maturity) are exploded, so much the better.

We can now note the following about the status of mathematics and Darwin's theory of evolution. First mathematics: (1) Much of mathematics enables us to use our other scientific tools more effectively, more precisely, and so on. The development of mathematics (e.g., the calculus) has often been undertaken as part of the development of another scientific theory. Mathematics, then, often fulfills the *beta'* criterion (the articulation of natural theory to fit it to the world). It is partly this connection with the search for knowledge that warrants our calling mathematics a science. That the development of mathematics has at its heart the development of the *beta'* criterion, can be seen from Galileo's demand for the use of a mathematical language in physical science and for Newton's motivation in incorporating such demand in his methodology. They both felt that without mathematical articulation, physics and astronomy could not go far. (2) The curiosity aspect which mathematics (and logic) share with the physical sciences constitutes another partial justification.[14] (3) Whether a particular discipline is called a science or not is to some extent a historical matter.[15] The beginning of mathematics was connected with the search for the nature of reality (mathematics gave us the structure of the world, according to the Pythagoreans). We may not agree with the Pythagoreans any more, but we can see that the 'function' of mathematics was thought to be what some people today think is the 'function' of the physical sciences.

Darwin's theory of evolution fits the *gamma'* criterion beautifully. Its main merit lies in enabling us to see that living creatures are the result of a history throughout which causal mechanisms have been in operation. Darwin showed us that novelties have come about, step by step, not due to Lamarckian historical tendencies (which kept natural history outside the realm of science) but because of the appropriateness to the situations encountered by the different organisms. Even though we cannot use the thesis of selective perpetuation to predict the future, Darwin provided a restructuring of much scientific knowledge which did lead to expectations very different from those of his predecessors: about the time scale necessary for life to have evolved (far larger than had been thought), about the continuity in the fossil data (which had the Darwinists on the defensive for a long time),[16] on several other geological matters, on some embryological concerns, and so on.

A more detailed account of how Darwin's Theory of Evolution satisfies the *gamma'* criterion, thus giving a new perspective to the question of the theory's scientific character, can be a fruitful undertaking.[17] It would surely include the manner in which it has paved the way for the birth and development of ethology and other areas of biology. But for the purpose of this essay it is sufficient to show that the *performance* (understanding) model of scientific knowledge need not keep us from ascribing the title of 'science' to several theories we consider scientific. We might have simply taken the bull by the horns and declared that the troublesome (or borderline) cases were not scientific because they did not meet our model.[18] Instead we may even shed some light on recent disputes by making us of the understanding criteria extrapolated from Scriven's account.

5.2 The value of science

What we have in mind by 'getting along in the universe' is that a species be able to survive, avoid great hardships, better the lot of its members, and so forth. As a result of this line of thought it was important to show that survival value was not incompatible with the function of science. Three points are not entirely clear yet, however.

(1) The informal account of 'getting along' so far provided seems to involve more than reproductive efficiency. How close is the analogy to biology then? And how justified?

(2) My presentation deals with a possible strategy: diversification and flexibility. But in the animal world we also see examples of extremely rigid accommodations to the environment. If the environment does not change, rigidity turns out to be a very good strategy for survival. Now, what rules out such strategy for the survival of a species such as ours?

(3) In answering the charge made in (2) I may argue that the strategy I advocate is needed for science to develop. But are the survival advantages of science all that clear anyhow? Many people think nowadays that the greatest danger to the survival of our species is precisely the technology derived from our science (nuclear weapons, etc.). Might not a species be better off with no science (or with a less advanced science)? Might it not last longer in such a case?

Response to (1). How much does a performance model need depend on any particular evolutionary approach in biology? One could perhaps manage in the style of John S. Mill, who said that 'even progress . . . for the most part only substitutes one partial and incomplete truth for another; improvement consisting chiefly in this, that the new fragment of truth is more wanted, more *adapted to the needs of the time than that which it displaces*'.[19] But it would be foolish not to explore the possible connection with a full-fledged theory of adaptation, an advantage Mill did not have. And in any event survival is a minimum test of 'getting along', and so in the last analysis evolutionary concerns are unavoidable for the performance required in this essay. Furthermore, the model must be in accordance with the biological considerations implicit in the relativism presented in chapter 3. That the account involves more than reproductive efficiency is not a great shortcoming, in my opinion. Even in the 'hard core' biological area there is more to adaptive value than producing a higher rate of offspring. For example, as E. O. Wilson says, 'For a genotype of a species that lives in a stable habitat, there is no Darwinian advantage to making a heavy commitment to reproduction if the effort reduces the chance of individual survival'.[20] There is room then for the additional considerations brought up in my account of 'getting along'. Even while undergoing change the environment still allows for long periods of relative stability, which permits a finer 'tuning' to a particular 'niche' so to speak. Thus a science structured to fit my performance model would not only create new opportunities for endeavor (dealing with new environments, and so on) but also suggest more efficient means for taking advantage of those opportunities already present. In this manner science may contribute to the variety and richness of life. Of course, some results of these contributions, some goals we may now pursue might not be biological in character; but they are in consonance with an overtly biological scheme of scientific knowledge.

Response to (2). According to Lorenz, we are 'specialists in non-specialization'. If so, we already have a hint as to why the strategy of diversification is to be favored. The environment does change, and it changes in many ways, some of which are of particular significance to the purposes of this essay. Its features may not remain constant (suppose, for example, that the stability of the sun in only temporary, a purely random state of relatively short duration).[21] Or its features come to be perceived in such a different way that the species finds itself in a completely different environment for all practical purposes. Or the species is constantly meeting new features of the universe. Of course, many of these sorts of changes may be combined. As an illustration one may think of the changes in the environment of a blind man who gains sight, and then one may recognize certain similarities in the development of cosmology in the last one hundred years. Having at our disposal many new experimental and theoretical tools, such as radio astronomy, nuclear physics and the theory of relativity, our picture of the universe has been so transformed that we might say it has exploded. It is as if the Copernican revolution had expanded with relativistic speed: the majestic sun went from the center of the universe to the suburbs of a rather big though still ordinary galaxy, and its majesty was replaced by the status of a very average, second-generation yellow star. We started with one galaxy, the Milky Way, and ended up observing billions, in just a few decades.[22] But neither the increase in the perceived size (time) of the universe nor the character of its fantastic newly-found inhabitants (neutron stars, quasars, black holes, etc.) is the main novelty in cosmology. No, even greater novelties are: the four dimensional nature of the universe, the unavoidable facing of choices between alternative non-Euclidean geometries, the consideration of whether the universe is open or closed (i.e. of whether gravity will ever stop its expansion.[23] Not only do we 'perceive' our environment anew, then, but must also confront so many wondrous aspects of it that were beyond even our imagination not long ago.

The way for more changes is being prepared all the time. For example, in determining the possibility of extraterrestial life we should consider what other biologies are possible, and we should also advance ideas that clearly do not spring from the requirements our science of biology so far has pressed upon us.[24] Thus we prepare ourselves to deal with areas that may barely keep a family resemblance to what we today consider biology, as well as to restructure radically our notions about the nature of life in our own planet. Such a new manner of viewing the universe would place us in a new environment, so to speak (would enable us to deal with a new variety of 'niches'). The general point to be made is that when talking about understanding we spoke of the ability to deal with a changing environment. Whether the 'actual' responsibility for the transformation in any given case lies with the environment, our science, or both, the result is much the same.

Even if this dynamic picture were to freeze, so that it were no longer reasonable to suppose that one must deal with a changing environment (and how could such a thing come about?), the possibility of change would still be there, and thus the rational need of preparing for it. The role of the mechanism for the generation of alternatives might then be greatly diminished, one may think. But it must be realized that nothing guarantees that the adaptation to a so-called static environment is as good as it can be, i.e. that the standard view is perfect. If so, then it is the task of science to present different views which, if fruitful, may lead to an even better adaptation. Thus we are led to a situation of change again.

Response to (3). I do not claim that science *guarantees* the survival of any intelligent species. I use expressions like 'enables to', 'tends to', and so on. A society may fail to take advantage of what its science discovers. Indeed, knowledge of the approach of a hurricane need not make the inhabitants of a region take the precautions they should. The following example may help to illustrate my position.

Suppose a group of people shipwrecks on a deserted island and, after a few years, loses all hope of rescue. Now let us imagine that one member of the group has the sort of knowledge of seismology that our society will probably have in the next century. This man is able to predict that an extremely powerful and highly localized earthquake will destroy the island in about a year. Through him the group learns what will happen on the island and may take measures to move to another island. Such steps will include the building of not-too-safe rafts and the braving of treacherous seas. These people's carpentry and navigational skills may be such that their chances of drowning are quite good. Because of the scientist's understanding about the earth, they may embark on a very risky trip. And they may also be very unhappy from the time the seismologist announces his findings.[25] Imagine now that no member of the group has any knowledge of seismology (or that the seismologist keeps the matter to himself). Instead of probably drowning in a few weeks (the trip, if any, must begin soon, for a long distance must be travelled), the group will live more or less comfortably for at least one more year. The chances of survival for that period are thus greatly enhanced by *not* having science (at least that particular science). But in the long run . . . [26]

This is the sort of point I want to make: the presence of science introduces risks we would not encounter otherwise (the production of nuclear weapons, for example). Thus in the short term science may or may not enable a species to 'get along' better (though I suspect a stronger case in favor of science might be made on this point too). Given a changing environment, however, the advantages science has to offer become indispensable over a very long period.[27] And, as we have already seen, such advantages go far beyond the merely indispensable.

References

1. Stephen Toulmin, *Foresight and Understanding*, Harper and Row, (New York), pp. 28–29.

2. Michael Scriven, 'The Concept of Comprehension: From Semantics to Software', *Language Comprehension and the Acquisition of Knowledge*, Roy A. Freedle and John B. Carroll, eds. V. H. Winstons and Sons, Inc. (1972), Carroll, p. 35.

3. I am not suggesting that Artificial Intelligence is doomed to failure.

4. O. Neugebauer, *The Exact Sciences in Antiquity*. Princeton University Press (152).

5. It would be interesting to determine whether Babylonian science exhibited to any degree the structural properties that, according to the social conception, science must have if it is to be rational. If one went by Toulmin's rendition, such properties were not present. Incidentally, at this point it is appropriate to suggest that much of the Greek science discussed was more an attempt at understanding than understanding itself according to the performance criteria.

6. For example, J. Bronowski in his television series (and book) *The Ascent of Man*.

7. K. R. Popper, *Objective Knowledge (An Evolutionary Approach)*, Oxford University Press (1972), pp. 242, 244.

8. K. Lorenz, *Studies in Animal Behaviour*, Harvard University Press (1971), p. 228.

9. *Ibid*., p. 228.

10. *Ibid*., p. 175.

11. *Ibid*., p. 175.

12. *Ibid*., p. 228.

13. *Ibid*., p. 234.

14. This would seem to be a purely intellectual exploration, but this supposition will be questioned later.

15. Of course, the important issue is whether it ought to be called a science. But then again, we should not put that much stake in demarcation criteria. Notice that the criteria I offer are *not* of demarcation, but rather of success in providing understanding on the part of theories, hypotheses, and the like. I would suspect that only some scientific theories are successful. Lack of success should not relegate the others to the unscientific pile. In general, any serious candidate to provide the sort of understanding discussed in these pages belongs in the fold of science. Many others will approach it to different degrees (I do not think there are sharp boundaries). How serious is serious enough is difficult to determine, but not all that crucial.

16. Until the publication of A. W. Roe's report on the fossil sea urchin *Micraster*.

17. I do not wish to suggest that no evolutionary theory (in biology) may also satisfy the *alpha'* criterion (by making accurate predictions, for example). Even though Darwin's own theory might have been too vague to permit a more direct confrontation with experience, its modern descendant, aided by the tools of molecular and population genetics, among others, may have far more to offer in that respect. Perhaps so. My examination does not require me to go beyond Darwin's position, however.

18. Or, as Popper and others do, separate logical (including mathematical) from other science. Reclassification alone does not solve epistemological problems.

19. John S. Mill, *On Liberty*, Bobbs-Merrill, (1956), p. 56. My italics.

20. Edward O. Wilson, *Sociobiology*, Harvard University Press (1975), p. 100. Wilson argues persuasively against certain standard interpretations, and in favor of a distinction between *r* and *K* selection. An *r* strategist is opportunistic (his approach is variable and flexible) in order to make 'use of a fluctuating environment and ephemeral resources' (p. 99). A high reproductive rate is important to *r* strategists. To *K* strategists (those that take advantage of a more stable environment) the concern with high reproductive rates is replaced by the need to 'maintain the densest populations at equilibrium'. For 'Genotypes less

able to survive and to reproduce under these long-term conditions of crowding will be eliminated', (p. 100). Considering only the immediate environment there might be an inclination to find in *homo sapiens* stronger K than r components (so to speak). But as one can see from the text below, the variability of the environment over long periods makes a strong does of r strategy highly advisable. (In respects to other than high reproductive rate. Although even this may become advantageous once again, e.g., if the colonization of the galaxy is ever feasible.) These last few remarks must be taken in the context of an analogy to evolutionary theory, not of an attempt to live by the letter of it.

21. The stability of the sun is in fact becoming a highly controversial matter in astronomy. Of course, if it turns out that the instability of the sun was only an illusion created by the state of our science, then I would not have an actual example. A more interesting case may be the decrease in the gravitational constant (as the universe expands) conjectured a decade or so ago.

22. It is important to note, however, that Kant had already claimed that some nebulae were galaxies over a hundred years before the subject became a crucial controversy in astronomy. See his *General History of Nature and Theory of the Heavens.*

23. If so, it will presumably contract and eventually die with a big bang, thus giving birth to another universe; if not, it will lose its energy and die with a whimper. According to the present evidence, T. S. Eliot's insight is true in the cosmic scale.

24. For example, Carl Sagan's suggestion that the appearance of life may be cyclical (e.g., in Mars).

25. This is to emphasize the disadvantages science brings to the community. We may also say that the group will fail r and k selections (in their extended, analogical senses).

26. Some may fear that science is too much out of phase with the other social structures of our culture. To illustrate this point suppose that doves quickly evolve a very powerful beak. (This is one of Lorenz's favorite examples.) At first one may think that the new beaks are invaluable in terms of survival. But given their scarcity of mechanisms against killing members of their own species, the effects of the new beaks would be devastating on any population of doves. But all this point shows is that certain cultural traits are maladaptive, since they block the long-run advantages of science. In any event, it fails to show that science does not offer such advantages. For a more specific discussion of this issue, see my 'The Moral Autonomy of Science and the Recombinant DNA Controversy', in the *Journal of Social and Biological Structures*, 2, (1979), pp. 235–243.

27. Carl Sagan suggested in *The Dragons of Eden*, Random House (1977), p. 230, that intelligence is adaptive up to the point where it produces science and technology of the sort we have today, but that its selective advantage is uncertain from that point on. I hope to have reduced that uncertainty.

6

EVOLUTIONARY EPISTEMOLOGY

In this chapter I will describe and criticize some views with which mine shares at least some slogans. I will be concerned with what has been alternatively called 'scientific epistemology', 'biological epistemology', and 'evolutionary epistemology'. Apart from bringing attention to very worthwhile work, I hope to show how my position avoids some mistakes that may plague views with similar motivations.[1]

The biological theory of knowledge had its modern beginning in Herbert Spencer's *The Principles of Psychology* (1855). Other early exponents of this sort of view were Avenarius, Baldwin, Bergson, William K. Clifford, Helmholtz, and especially Ernst Mach, who probably made the best case for it. Another figure who made valuable, though ignored, contributions was Henri Poincaré, better known for his conventionalism. In our century Konrad Lorenz's evolutionary epistemology constitutes a very impressive achievement. Jean Piaget's work can be considered of some relevance as well. Among professional philosophers, Popper and Toulmin have recently published ambitious attempts to provide an epistemology along these lines.

I would like to contrast the nineteenth and twentieth century approaches. The two differ in their accounts of scientific growth:

the older view ties science and survival value in an improper way, the present view draws analogies from neo-Darwinism but shies away from actual biological considerations. In spite of these shortcomings, I hope it will be seen nevertheless that great intellectual wealth can be found in both of them.

What is biological epistemology? I think Karl Popper captures the heart of the older version very well:

> . . . a sense organ is akin to a theory: according to evolutionist views a sense organ is developed in an attempt to adjust ourselves to the external world, to help us find the way through the world. *A scientific theory is an organ we develop outside our skin, while an organ is a theory we develop inside our skin.*[2] (my italics)

The functions of the mind are considered dynamic, that is, susceptible to growth and development, *at the level of the species.* What is being offered is simply a parallel between biological and psychological structures; just as the human being develops according to a certain mold, and attains a certain biological equilibrium, both of which are the results of a long evolutionary process, the human mind develops from infancy to maturity according to a certain mold and then settles into a more or less permanent functional structure,[3] both of which are also the results of a long evolutionary process.

Popper's insight can be taken in the sense that theories (in a strict understanding of theory, i.e., all-encompassing theories in some general field such as physics) provide us with a *cooperative* way of 'perceiving' the universe, thus they are like organs. It was in roughly these terms that the nineteenth century proponents of biological epistemology explained their view.

Spencer, at least, considered that our evolution was complete. His position amounted to a replacement of Kant's account of an a *priori* structure of the human mind by an evolutionary account which had as its final stage a neo-Kantian categorial ordering of the mind, viz. Newtonian, Euclidean, etc. The categories were the result of evolution, not of individual development. In Spencer's words, 'What is *a priori* for the individual is *a posteriori* for the species'.

The matter of the categories, that is of *which* categories, was hardly one of general agreement. Mach was a devastating critic of Newtonian physics, but felt that the structure of the human mind was Euclidean, and thought of logic (two-value logic) as an ideal limiting case of human thinking.[4] Nevertheless, such a structure was for him the result of an adaptive process; in speaking of the concept of causality, for example, he says that:

> Much of the authority of the ideas of cause and effect is due to the fact that they developed *instinctively* and involuntarily, and that we are distinctly sensible of having contributed nothing to their formation. We may, indeed, say that our sense of causality is not acquired by the individual, but has been perfected in the development of the race.[5]

Only what falls within the powers of the structure is conceivable. Thus inconceivability is a good indication that a proposal fails to be a genuine alternative, e.g., the inability to conceive of the world in non-Euclidean terms.[6]

Mach stressed the economical character of science. As he put it, 'Science is a minimal problem consisting of the completest possible presentment of facts with the least possible expenditure of thought'.[7] Science has the function of reproducing facts in thought in order to save, or replace, experiences. Through evolution the mind adapts itself to the world:

> It is not to be denied that many forms of thought were not originally acquired by the individual, but were antecedently formed, or rather prepared for, in the development of the species . . . [8]

As a result of such adaptation the mind comes to reflect the world (in loose terms 'the structure of the world becomes the structure of the mind').

Spencer's view of evolution was more Lamarckian than Darwinian. He felt that evolution had a direction, and a goal: man in his present and perfected state. As a result he thought that our evolution, including our psychological side, was completed. But it is not all that clear that Mach was all of one mind on the subject, as has been argued.[9] It is true that Mach said many things that would lead one to believe so (i.e., his remarks on inconceivability as a test). Nevertheless, there is a strain in Mach that points in the opposite direction, and which shows that he was rather ambivalent on the matter, and thus that he does not belong in Social-Darwinism as Spencer surely does.

Mach claimed that scientific thought is promoted by the gradual widening of our field of experience. This is not very controversial, but then he went on to say that 'the extension of our sphere of experience *always* involves a transformation of our ideas'[10] (my italics), and further that our sphere of action (of experience) was 'constantly widening'.[11] Thus the transformation of our ideas should be expected to continue. Of particular interest I find the last paragraph of his essay on mental adaptation:

> We are prepared, thus, to regard ourselves and every one of our ideas as a product and a subject of universal evolution; and in this way we shall advance sturdily and unimpeded along the paths which the future will throw open to us.[12]

His intent is made even clearer in the accompanying footnote:

> C. E. von Baer, the subsequent opponent of Darwin and Haeckel, has discussed in two beautiful addresses (*Das allgemeinste Gesetz der Natur in aller Entwickelung*, and *Welche Auffassung der lebenden Natur is die richtige, und wie ist diese Auffassung auf die Entomologie anzuwenden?*) the narrowness of the view which regards an animal in its existing state as finished and complete, instead of conceiving it as a phase in the series of evolutionary forms and regarding the species itself as a phase of the development of the animal world in general.[13]

Perhaps a mistake of interpretation of another aspect of Mach's philosophy would lead one to believe that Mach shared Spencer's view uncritically. According to Mach the basis of physical knowledge is provided by what he called 'psychological elements'. This position has normally been taken as a sense-data theory, which makes Mach, given his positivism, a precursor of the Vienna Circle. Many of his claims give credence to such an interpretation, e.g. 'a *thing* is a thought-symbol for a compound sensation of relative fixedness'.[14] Since with instrumentalism, a form of conventionalism, comes a conservative outlook about the solidity of the observation language (which in Mach's case would be equivalent to a language based on his 'elements'), it could be thought most plausible that Mach would be wary of truly radical alternatives, and would thus participate in the general nineteenth century belief about the completeness of our species' mental development. As it turns out, Mach regarded his 'elements' not as solid building blocks of knowledge but as highly theoretical. In fact, Mach thought that without theory experience would be impossible: 'Experience alone, without the ideas that are associated with it (theory), would forever remain strange to us'.[15]

After this short digression we can better evaluate Mach's admittedly unclear position. Over-generously perhaps we can say that even though Mach did think that the *present* structure of the mind had some of the same 'categories' that others insisted on, i.e. that it was Euclidean in character, he also suspected that our evolution was by no means complete, or at least, that there was no good reason to think that it was.

Now, for Mach there comes to be a correspondence between the structure of the mind and the structure of the world. Extending this point somewhat, we could say that we have *accepted* the all-encompassing theories we have, which constitute ways of cooperatively perceiving the world, *because they are true* (or if the evolution is not complete, because they are nearly true, or on the way to the truth, etc.). This position would be amenable to most nineteenth century epistemologists, but it need not be a crucial claim of the general view. This can be seen by contrasting Henri Poincaré's approach to Mach's.

Poincaré felt that we 'choose' such theories not because they are *true* but because they are more *convenient*. Since Poincaré is known as a conventionalist, most commentators take him to mean 'logically convenient' (or 'simpler', 'more elegant', etc.). It is clear that he often means that, but he also had a notion of *biologically convenient*:

> It has often been said that if individual experience could not create geometry the same is not true of ancestral experience. But what does it mean? Is it meant that we could not experimentally demonstrate Euclid's postulate, but that our ancestors have been able to do it? Not in the least. It is meant that by natural selection our mind had *adapted* itself to the conditions of the external world, that it has adopted the geometry most advantageous to the species; or in other words the *most convenient*. This is entirely in conformity with our conclusions; geometry is not true, it is advantageous.[16]

Poincaré was an associationist, it seems. Thus his biological epistemology took on an associationist character:

> It is this complex system of associations, it is this table of distributions, so to speak, which is all our geometry or, if you wish, all in our geometry that is instinctive. What we call our intuition of the straight line or of a distance is the consciousness of these associations and of the imperious character itself . . . An association will seem to us by so much the more indestructible as it is more ancient. But these associations are not, for the most part, conquest of the race. Natural selection had to bring about these conquests by so much the more quickly as they were the more necessary . . . We see to what depths of the unconscious we must descend to find the first traces of these spatial associations, since only the inferior parts of the nervous system are involved. Why be astonished then at the resistance we oppose to every attempt made to dissociate what so long has been associated? Now, it is *just this persistence that we call the evidence for the geometric truths*; this evidence is nothing but the repugnance we feel toward breaking with very old habits which have always proved good.[17]

Poincaré also provides a very nice synthesis to the rationalist-empiricist controversy:

> We see that if geometry is not an experimental science, it is science born apropos of experience; that we have created the space it studies, but adapting it to the world wherein we live. We have selected that most convenient space, but experience has guided our choice; as this choice has been unconscious we think it has been imposed on us; some say experience imposes it; others that we are born with our space ready made; we see from these preceding considerations what in these two opinions is the part of truth, what of error.[18]

In this Poincaré (as well as Mach) improves on Kant. Whereas Kant's solution suffered from the static character of his axiomatic system, the dynamic neo-Kantianism of the evolutionary epistemologists offered an alternative that I find at the very least interesting. The categories could be thought appropriate because of their adaptive value, because it could be shown (potentially) that they were *convenient* by showing that they resulted from evolutionary processes. The manner of their *genesis* then throws light on the question of their *justification*, philosophers' warnings about the genetic fallacy notwithstanding.[19] Perhaps such considerations should not be advanced in ethics, but they do not seem obviously out of place in epistemology.

Most of the great philosophers of the classical and modern periods can be considered philosophers of science (Plato, Aristotle, Descartes, Berkeley, Hume, Kant, and so on). What they did is the equivalent of what has been called philosophy of science in this century, even though we place their work under the headings of epistemology and metaphysics. Once outside the context of intellectual struggle in which it grew, their work may seem strange and idle to some. But it is not likely to seem as strange and idle as philosophical work that does not even benefit from similar beginnings. Since after Kant philosophy of science left the center stage of the discipline, it is perhaps not surprising that today's historical evaluations of the nineteenth century are rather unkind. Nonetheless, if we realize that the outstanding problems in philosophy (and I would like to say that Kant's fit the bill) were treated in a very interesting manner during that century, then a historical re-evaluation might be in order. And the considerations advanced in this chapter surely indicate that the work of the evolutionary epistemologists is interesting in precisely the relevant way. It is not unreasonable then to think of those philosophers of science (they certainly were that) as being in the mainstream of philosophy, as being the true heirs of Kant, for they updated Kantianism by developing it in the light of the most important idea of their century: evolution.[20]

Piaget's work has corroborated many of the insightful conjectures of the nineteenth century, such as the close connection between the mechanisms of perception and intelligence. According to Piaget science is an extension of intelligence, and intelligence is the form of equilibrium toward which all the structures arising out of perception, habit, and sensori-motor mechanisms tend. Logic would be the formalization of such a state of equilibrium (compare to Mach's claim that logic was an ideal limiting case). Since it would be extremely difficult to do any justice whatever to Piaget's voluminous and complex researches here, I must limit myself to a few remarks that shed some light on the general sort of position held by Mach and Poincaré.

Thanks to Piaget, it is easier to understand why our mental structure fails beyond the range of middle range objects: it is simply that such a structure is the result of adaptation to the middle range, between the microscopic and the megascopic. In quantum physics, for example, the most important 'category' of the middle range, the permanent identifiable object, is out of place. This gives us a perspective in the evaluation of Kant. As Milik Čapek puts it:

> What Kant analyzed was not the *a priori* structure of mind but its particular modification which is imposed on it by our continuous intercourse with the solid bodies of our macroscopic experience which eventually was systematized in the Euclid-Newtonian framework.[21]

It seems, then, that the range of our experience has been a limited one. Thus, following Mach's dictum, as that range varies and increases, our mental structure will change, perhaps completely beyond comprehension by our present intellectual powers. The extension of our field of experience need not be restricted to megascopic and microscopic phenomena. The exploration of space, and with it the exposure to environments in which the closely knit connections between the many biological and psychological functions may fail to operate, will offer a chance to study hitherto undreamed of perceptual and intellectual shortcomings and capabilities.[22]

It is not clear that Piaget would be altogether happy with Čapek's characterization of his view. Konrad Lorenz, however, is very clear on this matter: man's cognitive organs and functions have developed so that they work tolerably well within the 'middle ranges of their application'.[23] This is obvious in the case of perception in which:

> . . . the organization of peripheral and central receptor mechanisms is extremely 'narrow-minded' in its concentration on the practical requirements of survival of the species, and that it arbitrarily selects from reality

> only a restricted segment, which is just sufficient to meet these requirements and thus produces a 'twisted' picture of reality.[24]

Lorenz makes the same point about the 'higher' cognitive functions:

> (the) forms and categories of conceptualization are also functions of central nervous organization, which bear an analogy-relationship to the inherent reality of objects which is just as incomplete as the relationship between the colour 'red' and the electromagnetic waves of a given range of wavelengths.[25]

Lorenz's view is that our cognitive apparatus, from perception to the higher forms of intelligence, is the result of a long phylogenetic response to man's environment, but it is also a highly selective and thus limited response. It is in a sense a 'hypothesis' about the universe ('designed' to enable us to cope with the universe), a hypothesis put forth and developed in the history of the species. A position developed along these lines he calls 'hypothetical realism'. According to Lorenz a hypothetical realist 'does not expect (the cognitive functions) to convey, to us, absolute truths about outer reality, but just that kind of working knowledge which is necessary for the survival of the species'.[26] As examples of hypothetical realists he mentions physicists, particularly the pioneers of atomic physics, and biologists.[27]

It seems then that in some aspects Lorenz's view closely resembles Poincaré's. He also meets Mach on ground common to many evolutionary epistemologists:

> . . . it is not the *a priori* schematism of our conceptual processes and thought which arbitarily and independently prescribes for extra-subjective reality the form it assumes in our phenomenal world. In terms of phylogeny, it was the other way round: extra-subjective reality, in the course of aeons of persistent struggle for survival, has forced our developing 'world-image apparatus' to give due reckoning to its properties.[28]

But just where does science fit in all of this? We have been talking about the higher forms of thought. Are scientific theories to be included among those? Wouldn't it be more plausible to suppose that our higher forms of thought, our mental 'categories', constitute the limits within which our scientific theories can operate? If that is the case, Mach may have been mistaken in suggesting that scientific theories were the result of natural selection. The structure of the human mind might be, and that would be a result important enough for epistemology. But specific scientific theories would not, no matter how general their character. Certainly our biological make-up could not have possibly changed as much as, and as quickly as, our scientific outlook in the past few hundred years.

In its Machian version, evolutionary epistemology seems untenable (on biological grounds if nothing else), but it also seems that one can be a biological epistemologist without subscribing to what Stephen Toulmin has called the Machian fallacy. One way to do this is to sever the relation between scientific theories and survival value altogether. Such has been the tack taken by Popper and Toulmin, a tack that I find mistaken.

Popper has much in common with Mach. He thinks that science (or its precursor at any rate) began as a response to environmental pressure, and talks of science as a means of control.[29] Popper wants to distinguish animal from human evolution.

> *Animal evolution* proceeds largely, though not exclusively, by the modifications of organs (or behavior) or the emergence of new organs (or behavior). *Human evolution* proceeds, largely, by developing new organs *outside our bodies or persons*: 'exo-somatically', as biologists call it, or 'extra-personally'. These new organs are tools, or weapons, or machines, or houses.[30]

The great advantage of exosomatic evolution is that we let our theories die in our stead. If this is the case, then clearly there is a connection between science and survival value. But apparently he finds this position a bit too crude. So he amends his view in a way that I do not find very consistent. Suddenly he announces that he wants to offer 'something like a refutation of the now so fashionable view that human knowledge can only be understood as an instrument in our struggle for survival'.[31] He disclaims any such interpretation of his position.

> . . . I did not state that the fittest hypothesis is always the one which helps our own survival. I said, rather, that the fittest hypothesis is the one which best solves the *problem* it was designed to solve, and which resists criticism better than competing hypotheses.[32]

The suggestion seems to be that the adaptation of theories Popper has in mind is adaptation to an 'objective realm' separate from the world of things (i.e., from the universe), that is, adaptation to his 'Third World'. I find this matter of the Third World a most unhelpful form of neo-Platonism.[33] Nevertheless what is of interest in Popper's change of mind can also be found in Toulmin's *Human Understanding.*[34]

Toulmin claims against Mach that the tasks of scientific disciplines are intellectual, not biological or economic. According to him:

> Mach's account of scientific evolution shipwrecked chiefly because he equated the intellectual selection-criteria of science with the quite different criteria operative in organic change and economic development: viz., differential reproduction rate and productive efficiency.[35]

Toulmin attempts to provide a more general account of intellectual evolution. It is very difficult to do justice to his impressive (and certainly extensive) book in these pages. In essence, though, he thinks that a theory should be preferred if it adapts best to the *intellectual environment* it faces. The emphasis will be on how its performance measures against that of his rivals, given the problems that the scientific community of the time finds pressing and important. Toulmin demands mechanisms for variation and selective perpetuation (proliferation of alternatives plus what Feyerabend calls the 'principle of tenacity'). For these mechanisms to operate, however, there must be a forum or a court in which the new alternatives may be heard, and a tribunal that will preserve the accepted view until one of the alternatives convinces it that it is better adapted (or perhaps, adaptable) to the discipline's intellectual environment (problems, ideals, etc.). This tribunal is rather formally constituted and it requires a professional society for its existence.

Toulmin places great emphasis on the existence of such tribunals as the main element that would differentiate scientific from other disciplines. This move, he feels, allows him to escape the Machian fallacy, which must come as a great relief to him since he thinks that 'except in the most flatulently extended sense of the words, the scientist cannot be said to value his ideas on account of their "biological" or "economic" advantages'.[36] Toulmin's position, then, is 'evolutionary' or 'biological' only in an analogical sense.

Toulmin tries too hard to avoid this Machian mistake. As a result he fails to draw as sharp a distinction as he would like between scientific and non-scientific intellectual disciplines. It seems that in his view what keeps ethics and art from being sciences is mainly the lack of professional societies which would provide the forum and the tribunal for the proposal and elimination of alternatives.[37] But surely more is needed. As Toulmin very well knows, such professional societies have existed throughout the history of art. The French Academy, for example, had a very rigid notion of what painting was. Alternative views of painting had long struggles in their quest for recognition. A crucial part of their struggle can be fairly characterized as a claim to the effect that their approach to painting is better adapted to the cultural environment (both the Romantics and Impressionists wanted to change the function of painting vs. the general culture; they both felt that painting could and should capture much more of man's experience).

Even if my historical characterization is not accepted, the point is that we can easily imagine a situation which will seem to make art fit Toulmin's bill. Suppose, for example, that for something to be taken as sculpture by the art profession (of which there would be a society and an academy) it would have to be a three-dimensional work made in bronze or marble (wood wouldn't be permanent enough) and according to certain methods (with a chisel, for instance). Now somebody decides that materials made easily available by the new technology, such as aluminum and hard plastics, can far extend the possibilities of a sculptor in bringing his art to the world.[38] It is not hard to see how a good case could be made for it. Nor is it hard to imagine that after some initial conservative reaction the professional society may switch to the view with the wider scope. This might well be an account of art as a rational activity. But it would be odd to say that art had thus become a science.

Scientific disciplines need not be (nor are they) our only rational enterprises. Some additional aspects must distinguish them from other activities (although I would not demand abrupt distinctions). Those aspects, I suggested in chapter 4, connect scientific theories to 'getting along' in the universe.

It is fair to say that the suspicion against the connection between scientific merit and survival value (and ultimately 'getting along') is motivated by the notion that survival value involves immediate or foreseeable applications. One often hears that whereas animals can only react to pressing problems, we can pursue long-term endeavors which are free from any compelling demands of the environment. *We* act out of curiosity. Science is the result of exercising a higher form of this quality of ours. As Konrad Lorenz has shown, however, curiosity not only exists in animals such as the Norway rat and the raven,[39] but it is also of great survival value. Curiosity allows such species to 'construct' their envornments for themselves, that is, to exercise a great versatility of application due to their minor degree of specialization (they are equipped with very few and very broad releasing mechanisms and not many innate motor patterns).[40] As a result, such species, which Lorenz calls 'specialists in non-specialization', can adapt to a great variety of environments. This is partly the key to the survival value of curiosity.

In dealing with more formal objections to the connection at issue, I made use of the performance model of understanding. It was seen then that the *beta'* and *gamma'* criteria, which provide for the refinement or incorporation of theories of more direct application (potentially), constitute indirect but secure links between a species' scientific understanding and its universe. More specifically, it was seen how they contribute to dealing with greater ease with our environmental situation (our 'niche'), increasing the number and diversity of environments that we can deal with (enlarging the 'niche'), and coping with a continuously changing environment (which puts a premium on flexibility of response). This contribution is, again, indirect – 'survival at a distance', as I fancifully called it.

By keeping these criteria in mind one should be able to see that a compromise is possible between nineteenth and twentieth century evolutionary epistemologists. One view was too strong, the other too narrow in conceiving the nature of the connection between scientific views and survival value.

From what has been said so far it is evident that we must distinguish merely theoretical changes in science from changes of basic conceptual forms (whatever they are). Once this distinction is made we should wonder about the relationship between the two kinds of changes. These and other related matters will be discussed in the following two chapters.

References

1. But this chapter is not intended as a comprehensive summary, let alone history, of evolutionary epistemology. For such a summary the reader is better advised to consult Donald T. Campbell's 'Evolutionary Epistemology' in P. A. Schilpp (ed.), *The Philosophy of Karl Popper*, Vol. 14, I & II, *The Library of Living Philosophers*, pp. 413–63 Open Court, Ill., (1974). Campbell himself has made some very intriguing suggestions in this field.

2. Karl Popper, 'Is There an Epistemological Problem of Perception? ', in Imre Lakatos, Alan Musgrave, (eds.), *Problems in the Philosophy of Science*, North-Holland Publishing Company, Amsterdam (1968), p. 163. Here Popper is presenting it as his own view. But we will see below that he is not of one mind on the subject.

3. The sort of account that Jean Piaget has given. See, for example, his Psychology of Intelligence, Littlefield, Adams and Co., New Jersey (1972).

4. Mach's biological theory of knowledge is best summarized in his essays 'On the Economical Nature of Physical Inquiry', and 'On Transformation and Adaptation of Scientific Thought', both of which appeared in his *Popular Scientific Lectures*, Open Court (1943), pp. 186–235 (future references to this work will appear under the heading *Lectures*), and in section IV of chapter IV of his *Science of Mechanics*, Open Court (1942), p. 582 (future references to this work will appear under the heading *SM*). Some may think that Mach's praise of Riemann indicates that he did not think the world must be Euclidean. This is incorrect. From 1901 to 1903 Mach published in *The Monist* the essays which he later collected in his *Space and Geometry*. In the Open Court edition (1943), p. 135 (a section appropriately entitled 'Applicability of the Different

Systems to Reality'), Mach states:

> Analogues of the geometry we are familiar with are constructed of broader and more general assumptions for any number of dimensions, with no pretension of being regarded as more than intellectual scientific experiments and with no idea of being applied to reality. In support of my remark it will be sufficient to advert to the advances made in mathematics by Clifford, Klein, Lie, and others. Seldom have thinkers become so absorbed in revery, or so far estranged from reality, as to imagine for our space a number of dimensions *exceeding the three of the given space of sense*, or to conceive of representing that space by any geometry that departs appreciably from the Euclidean. Gauss, Lobachévski, Bolyai, and Riemann were perfectly clear on this point, and cannot certainly be held responsible for the grotesque fictions which were subsequently constructed in this domain.

5. *SM*, p. 582.

6. In this I have been influenced by Milic Čapek, who in his essay 'Ernst Mach's Biological Theory of Knowledge' remarks that

> Having, like Spencer, defined truth and error in the terms of success and failure, Mach simply could not believe that the belief which survived a practically limitless number of tests can ever in the future be revised or modified. The belief in causality in its precise form of the law of constancy of energy was for him that uniformity of thought which reflected the most general and most pervasive feature of nature – its constancy and uniformity.
>
> *This was the reason why Mach, again like Spencer, regarded the inconceivability of the opposites as the most reliable epistemological criterion.* (my italics)

Boston Studies in the Philosophy of Science, Vol. 5, pp. 408–09.

7. *SM*, p. 587.

8. *Lectures*, p. 222.

9. Milic Čapek, *ibid.*, pp. 400–20.

10. *Lectures*, p. 229.

11. *Lectures*, p. 233.

12. *Lectures*, p. 235.

13. *Lectures*, p. 235.

14. *SM*, p. 580.

15. *SM*, p. 588.

16. H. Poincaré, *The Foundations of Science*, The Science Press, Lancaster (1946), p. 91.

17. *Ibid.*, pp. 420–21.

18. *Ibid.*, p. 428.

19. They would be justified insofar as they are generally appropriate or convenient in a *limited* domain. They may also be, in a sense, logically prior to experience; but as Popper points out, this is far from saying that they are valid *a priori*.

20. Of course, they were not recognized by their contemporaries as great figures (not that their work was generally considered worthless), but then neither were Berkeley or Hume by theirs. When I speak of contemporary historical evaluations I should specify that I mean analytic ones. The appraisal one findds in the European continent is quite different. It is not altogether unfair to claim that the bad reviews given the past century must be taken with a grain of salt, for the analytic approach is very a-historical, if not downright

anti-historical. In my view both the continental and analytic approaches are mistaken.

21. Milic Čapek, 'The Significance of Piaget's Researches to the Psychogenesis of Atomism', *Boston Studies in the Philosophy of Science*, Vol. 8, pp. 446–55.

22. See the last chapter of R. L. Gregory's *Eye and Brain*, McGraw-Hill (1966), pp. 229–39.

23. Konrad Lorenz, *Studies in Animal Behavior*, Harvard University Press (1971), p. 261.

24. *Ibid.*, p. 286.

25. *Ibid.*, p. 287.

26. *Ibid.*, p. 253.

27. '(Atomic physicists) are fully aware that our *a priori* forms of ideation and of thought, inevitable and logical though they seem, are nothing more than working hypotheses which fit the facts of reality no better than man-made ones do and which are accessible to constructive criticism in exactly the same way.' Lorenz, *ibid.*, p. 259.

28. *Ibid.*, p. 289.

29. Karl R. Popper, *Objective Knowledge*, Oxford (1972), pp. 239–40.

30. *Ibid.*, p. 238.

31. *Ibid.*, p. 264.

32. *Ibid.*

33. For a criticism of Popper's view I refer the reader to Paul Feyerabend's review. *Inquiry*, 17, January, 1975.

34. Stephen Toulmin, *Human Understanding*, Vol. I, Princeton University Press (1972).

35. *Ibid.*, p. 321.

36. *Ibid.*, p. 320.

37. In this discussion I do not wish to deny that ethics *could* be a science (i.e., applied social science) nor that there is a connection between art and the search for knowledge.

38. It wouldn't be just a matter of materials, but of the new possibilities in form, distribution of space, dynamism, etc.

39. Lorenz, *op. cit.*, p. 228.

40. *Ibid.*, p. 175.

7

THE LIMITS OF KNOWLEDGE

There are two important things we want to learn about knowledge: its nature and its limits. It is now time to turn to the second of these concerns. Is Total Knowledge possible? To answer this question one must first cross a field mined with platitudes about the shortcomings of finite beings. It is best to render any likely explosions harmless by remembering that the performance model of knowledge requires neither perfection nor infinity. Safe on the other side, one must realize that even if Total Knowledge is possible (for some being or other), it need not be within *human* reach. The question then stands amended: Is Total Knowledge possible for us?

It is most implausible to suggest that our present science already amounts to cosmological 'omniscience'. In fact, it is implausible to suggest that we are not very far from such a state. If so, the attainment of Total Knowledge may well require fundamental, radical changes in our science, in our very modes of conceptualization. But are any changes of that sort possible? Much of contemporary philosophy rules them out. I will argue that contemporary philosophy is mistaken.

The issue is whether there are any limits to the extent and manner in which scientific knowledge may grow. Some philosophers think there are. I will object to the two main positions they advance. The first one is an (conceptual) argument against the possibility of radical change. I will claim that crucial empirical assumptions are made in what purports to be a pure conceptual investigation. The second one is exemplified by Lakatos' methodological demand that the scientific change be gradual (a characterization he would have disliked, but a fair one as I hope to show).

Analytic philosophy has, for a long time, insisted in a distinction between 'empirical' problems and 'philosophical' (i.e. 'conceptual') problems. Questions such as 'What can we know? ', 'what sorts of things can we know? ', 'how much can we know? ', 'what are the limits of human knowledge? ' are practically paradigm epistemological questions, and thus they should be 'philosophical' in the analytic sense.

It is precisely in dealing with questions of this sort, however, that one can cast doubt upon the legitimacy of the analytic distinction. The crucial considerations in determining the limits of knowledge will turn out to be partly 'empirical'. (By 'empirical' we should not take only experimental results and the like, though; for a lot of highly theoretical concerns are always essential in structuring, guiding, and interpreting research.)[1] My position in the first part of this chapter is not that no limits can be fixed, but rather that no limits can be fixed by conceptual means alone. My point will be about philosophical methodology, then. Any claims I wish to make beyond this will be taken up in the following chapter. This task, by the way, does not coincide with W. V. Quine's attempt to deny the distinction between analytic and synthetic statements. Nor are its goals on a par with that philosopher's intent in 'naturalizing' epistemology. As the 'conceptual' argument I will discuss shortly seems to indicate, Quine would not be adverse to the standard analytic approach in settling such questions.[2]

The philosopher is supposed to establish the structure of our conceptual scheme. For by 'mapping out' such a scheme he learns 'how much we can know, 'what we can know', and so on. That is, he arrives at the limits, or outer boundaries of our knowledge. Whatever distinctions, classifications, etc. he makes in the process will be 'logical' or 'linguistic' (in the analytic sense of 'linguistic'). Thus the problem of the limits of knowledge is a problem of conceptual schemes. It is a matter of dispute whether the boundaries in question will turn out to be the limits of logic, or whether more of a fine structure can be provided, e.g. that some elements of the conceptual system are essential, or basic, to the very *possibility* of empirical knowledge (the sort of thing undertaken by P. F. Strawson in *Individuals*, and by Wilfrid Sellars in several writings).

Analytic philosophy allows several approaches. The problem may be restated and a solution attempted in a formal, unambiguous language, or fine points may be made about the range of crucial terms like 'know', and so on, in ordinary language, or it may be 'discovered' that certain 'linguistic abuses' are committed in the posing of the questions, or what not. A common thread, however, is the thesis that no 'empirical' observations or considerations have any bearing on settling such questions. This thesis is of course restricted to the 'essential' features of the issue and excludes the 'accidental' ones. That is, it concerns us with what may or may not be known 'in principle', not with mere likely knowledge. I believe that the analytic philosopher must insist on this distinction, for failure to meet some rather trivial empirical requirements makes cosmological knowledge very unlikely.

There is a biological requirement. A dog could not understand the universe to the extent that we can, *simply because it does not have the brain to do it*. And it would not be surprising that some creatures were capable of understanding the universe even less than dogs, and that others might do so far more. Likewise some creatures may have perceptual equipment that helps them in exploring certain profitable cosmological avenues. Others may not. There is a whole host of other empirical considerations, i.e. meteorological, geological, astronomical, etc. Suppose that one other race of creatures is as well endowed biologically as we are. But as it turns out, their planet is permanently surrounded by a thick layer of clouds (or take the case of a race of deep sea water creatures, or that of a race of sightless *homo sapiens*). Thus they will never see the stars. The development of astronomy will be most unlikely. Or suppose their solar system is engulfed in a black interstellar cloud – their astronomy would be very limited.

These requirements, it would be argued, do not touch upon the heart of the philosophical, i.e. epistemological, concerns we have. That is, what we do in philosophy is in addition to pointing out the empirical considerations. After all, if our conceptual scheme is such and such, how can we have knowledge beyond its boundaries? The limits of the conceptual scheme must then be the limits within which our knowledge must grow.

This neo-Kantian position of analytic philosophy is more modest than others held by the tradition. It was thought for a time that by analyzing our concepts we would somehow arrive at the 'structure of reality'. Now the claim is that conceptual (or linguistic) analysis will give us the nature of the mental (or perhaps, the 'conceptual' restraints the mind has). From philosophy of cosmology we have retreated to philosophy of psychology. Can a stand be made here? No.

Suppose that we meet a race with a science far more powerful than ours. It seems that they have either achieved total knowledge or are very close to doing so (as far as they themselves can tell). And suppose further that we are trying to spread the news, part of which includes the claim that this super race has a multi-value logic. (The notion that a multi-value logic should be developed in order to understand quantum mechanics better has been proposed several times.) It is important to note once again that the nature of a conceptual scheme is not very clear. But be that as it may, it is obvious that all sides agree that elementary logic constitutes the 'outer boundaries' of whatever turns out to be the conceptual scheme. What would the analytic reaction to this super-race case be?

My guess is that the case would be thought to make no sense. This is a consequence of the general analytic outlook, and can be traced in particular to Quine's famous doctrine of indeterminacy of translation.[3] Whether it is fair to impute to him the following argument is not as important as the fact that such argument crystallizes a position that lurks behind a certain way of doing philosophy. (This argument is, of course, one version of a sort that is generally advanced against diverse cases of possible radical conceptual changes).

(a) For the alleged alternative (the multi-value logic the super race is supposed to employ) to be an actual alternative it must be made intelligible.
(b) But to be made intelligible it must be *translated* (by means of analytical hypotheses) into a language we understand, i.e. English. (It must be stated within our conceptual scheme.)
(c) But if it is made intelligible in our language, then we already have those alleged extra values in our language. (It has been enclosed within the two-value logic of our conceptual scheme. Thus there are only two values after all.)[4]

Therefore:

(d) If the alleged alternative can be made intelligible, then it is not an alternative after all. And if it cannot be made intelligible, then the proposal cannot even get off the ground.

There are many things wrong with an argument of this sort. I will mention a few that have bearing on the main topic.

(1) It is not the case that for an alternative to be made intelligible it must be translated into a language we already understand (b). Take the case of a very primitive bushman whose language is extremely poor (conceptually). Suddenly he is exposed to a 'civilized' language (of an extreme conceptual richness). In a few years he manages to master the 'civilized' language. Has he made analytical hypothesis? Has he translated the very complex conceptual structures of the 'civilized' people into his own backward, primitive, conceptually poor language in order to understand them? But then he could not have understood them, for his language, *ex hypothesi*, did not have any counterparts. Nevertheless, he does understand them. Thus he must have come by the alternative without translation. He did so by learning the new language just as he did his first one: by sharing in the way of life. Perhaps one could make some headway into another language by means of analytical hypotheses. But the notion that one *must* use them sounds preposterous. Even in the case of a field linguist (whose language is, *ex hypothesi*, conceptually richer) it is just not the case that translation is going on. Paul Feyerabend, in developing a similar objection, pointed out that the field linguist learns the language precisely by sharing in the life of the community where the language is spoken.[5] Besides, how would a child learn his language? Certainly not by making analytical hypotheses and then proceeding to translation. What does he have to translate into?[6] What argument, what evidence at all, do analytic philosophers, or anyone else, have to *conclude* that a second language cannot *in principle* be learned like the first? None. This is a most gratuitous assumption.

It can be noted, of course, that humans have great difficulty in learning new languages past a certain age. Several hypotheses can be proposed to account for this situation, e.g. the dendrites of the cerebrum's neurons settle into rather rigid nets after a certain age. Thus there may be at least a large difference of degree between the way languages are learned in infancy, when the dendrite nets were yet to be made. But this and similar arguments are certainly 'empirical' on the one hand. (It must also be supplemented by further empirical hypotheses: that the process is irreversible, that the neural nets are permanently fixed, etc.) On the other hand it does not show, nor could it, that a second language could not be learned in infancy just like the first one.

(2) It could be argued that (1) does not meet the analytic argument, for both the bushman and the 'civilized' man share a common

basic conceptual structure. And the super race case requires that some humans go beyond their basic conceptual structure. But why couldn't the members of the super race teach their language to humans? Why is it that even if infant humans went to live in the super race communities not one of them could learn this super language? Because their brain would not measure up? Perhaps. But this is an 'empirical' consideration if any is! Even so, does the objection miss the point after all? Here we seem to come to the heart of the matter. In order to understand something we must think. In order to think we must think in some way. The way in which we think is structured by elementary logic, viz. two-value logic. If the two-value logic goes, then so does the way in which we think, and thus the very possibility of our thinking and understanding anything. This might be so. *But what is at issue is whether we can acquire a new way of thinking.* And surely, if we acquire a new way of thinking, as long as it is a way of thinking, we are again in a position which permits understanding. The points made in (1) with respect to the way the child learns his first language, and to what if any, restrictions it imposes on its learning a second language, apply here *mutatis mutandis*, unless further assumptions are made: (i) that our psychological apparatus is uniquely determined by our nervous system; (ii) that such psychological apparatus is forever fixed; (iii) that a two-value 'conceptual scheme' uniquely fits our psychological apparatus.

The plausibility of assumptions such as these will be examined in the following chapter. But I think it is fair to conclude that the boundaries of knowledge cannot be delineated by 'conceptual' investigations alone. The important considerations are crucially 'empirical'. (Both 'conceptual' and 'empirical' as understood by the analytic epistemologists.)

(3) It is claimed in (c) that if an (alleged) alternative can be made intelligible in our language, then our language contained the alternative prior to its being made intelligible. This need not be so. The home language may be *extended* to accommodate an alternative that other language (culture) presses on it. This is one of the main ways in which languages have changed under the influence of the language of the dominant culture of the time. Then the bushman of (1) could return to his native community and effect a series of changes in its language (that would amount to radical changes) such that the other bushmen can now *understand* the 'civilized' conceptual alternatives, even though they have not learned the 'civilized' language. Of course, the extension of the home language

will have at least a strong family resemblance (phonetically and otherwise) to the previous body of the language. One can examine the repetition of this process throughout the linguistic history of mankind.

(4) It is unreasonable to demand that in order to raise the possibility of an alternative, one *must* provide the alternative itself. Kant made this mistake in supposing that physics would never differ from Newtonian physics, that it could not change. And the position of Newtonian physics at the time was almost as well entrenched as that of two-value logic today. Of course, it may well happen that two-value logic is never overturned. For, after all, raising the possibility of an alternative does not provide the alternative itself.

Once again, as in chapter 3, some of the pillars on which crucial epistemological issues rest turn out to be 'empirical'. It does not follow from this, however, that a clear understanding of the nature of the relevant empirical considerations will determine the limits of knowledge. Several barriers to such optimism will be raised in the next chapter. And some complications will be mentioned immediately.

Throughout this paper I have been talking of *our* conceptual scheme, of *human* goals, etc. But just who is the *we* of whom it is considered whether they may or may not achieve Total Knowledge? It seems that there should be no difficulty in answering this question, for it is easy to tell *homo sapiens* from other species.

Nevertheless, answering the question is not as easy as it seems. Suppose we ask about a fifteen-year-old boy whether he is capable of attaining a level of intellectual achievement A, not right away but at some time during his life. It is hard to say, particularly if A is rather complicated. Suppose further that we ask the question about a ten-year-old boy? Or about a six-year-old? It is much harder to answer. Likewise with species. For all we know *homo sapiens* may be in their evolutionary infancy or adolescence. By the time the species matures it may be endowed with the appropriate brain and perceptual apparatus to attain Total Knowledge, given that the other circumstances are right. If the changes are substantial, however, one would have to worry about species criteria and the like. After all, if our highly evolved descendents attain Total Knowledge a hundred million years from now, it is not clear that the claim that *humans* could was correct.

The complications can multiply easily. The 'human' society of the future may be a mixture of *homo sapiens* and super computers (not limited to digital computers). If such a society achieves Total

Knowledge whereas a strictly human one could not have, is the claim correct? Suppose that an individual *P* suffers a special kind of cancer that attacks his organs one by one, tissue by tissue. Doctors replace his tissues, one by one, with synthetic tissues. This sort of replacement takes place today. But unlike today's patients, *P* turns out to have had all his tissues replaced with synthetic ones, over a very long period of time, and always one at a time. He still is *P*. Could not our species undergo a similar process (in that some of the human intellectual activities, e.g. science, are taken over by super-computers)? And were the species, so modified, to attain Total Knowledge, is the claim that *we* can correct?

It is plausible to expect the number and variety of puzzling difficulties to increase as our 'empirical' probe advances. But enough has been said on this topic for the *limited* purposes of this essay.

The second limitation to the growth of knowledge pertains not so much to content but to manner. It is said that the growth of knowledge *must* be *gradual.* This claim practically amounts to a truism when it simply expresses (a bit carelessly) the rather trivial observation that constructing a worthwhile alternative scientific theory is an enormous undertaking that requires much time and effort, by many people, sometimes by many generations of scientists. Positions are not usually razed to the ground but rather dismantled slowly, and sometimes their foundations serve as provisional foundations for those views that will replace them. There often is intellectual profit in extracting as much as possible from the older theory. Thus in certain cases it might be good methodological *advice* to use the old materials in constructing the new edifices of science.[7]

It is not uncommon, however, to take the 'must' literally. One then hears demands for 'continuity' or for 'cumulative' growth, demands so strong, furthermore, that failure to meet them seems tantamount to giving up on the rationality of science. But surely, the theses so far advanced in this essay make such demands appear very implausible. If radical change is permitted, the growth of science need not be as described above. And even if science pulls itself up by its bootstraps during the transition, there need not be anything in common between old and new when the task of building is completed. Such demands, then, amount to this: that some part of every replaced theory *must* be preserved in the theory that replaces it. But why should portions of empirical views be exempt from risks that even the most fundamental elements of our logical and mathematical systems must face? What justifies this step from advice to demand?

Of course it should not come as a surprise that the thinkers who advance this kind of demand also deny the possibility of radical change. Against them I would just repeat the points already argued.

There is one methodology, nonetheless, which sympathizes with the thesis that any part of science may be overthrown, but at the same time appears to require cumulative growth. Such is Imre Lakatos' methodology of scientific research programs. I proceed to discuss those aspects of it relevant to the present issue.

Imre Lakatos improved greatly on Karl Popper's methodological falsificationism, but there is a conservative strain in his philosophy of science that needs to be brought to light, criticized, and, if I am correct, expurgated. Instead of Popper's two-cornered fight between a theory and the 'facts', Lakatos correctly pointed out that crucial tests for a theory come about only in competition with another theory.[8] According to Lakatos:

> . . . a scientific theory T is *falsified* if and only if another theory T′ has been proposed with the following characteristics: (1) T′ has excess empirical content over T: that is, it predicts *novel* facts, that is facts improbable in the light of, or even forbidden, by T; (2) T′ explains the previous success of T, that is, all the *unrefuted* content of T is included (within the limits of observational error) in the content of T′; and (3) some of the excess content of T′ is corroborated.[9]

In pitting one theory against a rival, however, we normally elaborate and improve them both. In fact, we arrive at new theories. Thus what we appraise are *series* of theories rather than isolated theories. Such a series is constituted according to the *positive heuristic* of a scientific *research program.*

The positive heuristic provides a 'protective belt' around the *core* of the program. Both heuristics (there is a negative heuristic as well) generate a certain *continuity* among the theories of the series, for the negative heuristic tells us what paths of research to avoid, and the positive heuristic what paths to pursue. Lakatos gives criteria for progress and stagnation within a program.

> A research programme is said to be *progressing* as long as its theoretical growth anticipates its empirical growth, that is, as long as it keeps predicting novel facts with some success ('progressive problemshift'); it is *stagnating* if its theoretical growth lags behind its empirical growth, that is, as long as it gives only *post-hoc* explanations either of chance discoveries or of facts anticipated by, and discovered in, a rival programme ('*degenerating problemshift*').[10]

Within a program a theory can be eliminated only by the application of the 'falsification' (or replacement) criteria given above.

There is an objective reason to reject a program, that is, 'to eliminate its hard core and its program for constructing protective belts'. That objective reason is provided 'by a rival research program which explains the previous success of its rival and supersedes it by a further display of *heuristic power*'.[11]

Lakatos has soldered the idea of growth and empirical character into one. But his account of growth has been unfortunately limited by his second kind of requirement, namely that a new theory (or research program) must explain the *unrefuted* content of the theory (or research program) it replaces.[12] At first sight this requirement seems innocent enough. After all, we want our changes of theories to be content-increasing. Thus we exact of a new alternative as much as we did from the old one, plus some novel facts.

What is not considered, however, is that in every change of theory there are losses as well as gains.[13] I will not argue for the historical side of this thesis here. Instead, I will try to show that many important aspects of Lakatos' methodology supports this thesis against the conservative requirement. In fact, the requirement is *inconsistent* with what Lakatos saw as the main advantage of his methodology over Popper's.[14]

What could Lakatos mean by 'unrefuted content'? Certainly not 'unfalsified' or 'unrefuted' in Popper's sense. For, as Lakatos says, '... *no experiment, experimental report, observation statement or well-corroborated low-level falsifying hypothesis alone can lead to falsification. There is no falsification before the emergence of a better theory*'.[15] In Popper's sense a theory is 'refuted' when it clashes with an 'observation' statement (after certain precautions are taken); but then new theories are born 'refuted'.[16] A theory (or a research program) is refuted, in Lakatos' sense, only when it is replaced by a better theory (or research program).[17]

The unrefuted content of a theory T that has been replaced by another theory T′ is just that part of T that has been preserved in T′. It is 'unrefuted' as long as it remains 'unreplaced'.[18] But then the second requirement is trivial and pointless: 'preserve the portion of T that you preserve'.[19]

We could ascribe this difficult to a confusion on Lakatos' part, and amend the second requirement. What would be a suitable amendment?

We may want to preserve what T explained well. But this suggestion is too vague and beset with problems. The best formulation for the second requirement, and probably the one Lakatos was groping for, is the following:

(2) T′ explains the previous success of T, that is, all the well-corroborated content of T is included (within the limits of observational error) in the content of T′.

Not even in this amended form will the requirement do, however: (i) in the light of T′ the 'previous success' of T may seem an illusion; (ii) the content of T in question might have been 'well-corroborated' in accordance with an observation theory O which another observation theory O′, brought about by the emergence of T′, over-throws and replaces; in the light of O′ the content of T in question may no longer be considered 'well-corroborated'.

The point is that whether *any* part of T can be considered 'well corroborated' (or 'successfully explained') will depend on T′, not on the 'standard' of success or corroboration previous to the replacement of T by T′. Thus T′ *may* overthrow any part of T (or all of it) directly (i), or indirectly, by means of an associated observation theory O′ (ii).[20]

Lakatos himself provides excellent support for my position. According to Lakatos, the proponent of a theory (or a research program) that has been rejected can *appeal* the decision by challenging the relevant observation theory. He may take his own theory as an observation theory with which the interpretative theory (previously the observation theory) can be judged, or he may propose a new observation theory for the same purpose. If such a move results in a progressive shift, his own theory (or program) will be victorious, and the bothersome 'well-corroborated' observations will be overthrown.[21] Lakatos illustrated this point well with his account of the progress of Prout's program in a sea of anomalies.[22] (For example, Prout's program required that the atomic weights of elements be whole numbers, but the atomic weight of 'pure' chlorine, as measured in accordance with the very best experimental techniques of the time, was 35.6. This situation prevailed for the best part of a century until the development of atomic physics made it clear that the samples used contained two isotopes, Cl_{35} and Cl_{36}). A similar example could be provided by the triumph of the kinetic theory over the phenomenological theory of heat.[23] The fight need not be long, difficult, and/or indirect, though. As Lakatos puts it:

> Appeal procedures too are occasionally easy: in many cases the challenged observational theory, far from being well corroborated itself, is in fact an inarticulate, naive, 'hidden' assumption; it is only the challenge which reveals the existence of this hidden assumption, and brings about its articulation, testing, and downfall.[24]

Lakatos' admonition to regard properly the limits of observational error can be of little help against my objections, since, once again, such limits will be drawn in accordance to T′ or an associated O′. The replaced 'precise' observations may seem even silly in the light of T′, as Lakatos himself rejoiced in pointing out.[25]

Lakatos' methodology is supposed to reduce greatly the conventionalist element (and hence the arbitrariness) of Popper's methodology, by demanding that a theory be rejected only when progressively replaced by a rival. But the requirement for the preservation of the unrefuted content of a defeated theory can have any teeth in it only if we take 'unrefuted' in the Popperian sense (thus reintroducing the arbitrary element). In Lakatos' sense of 'unrefuted', the requirement is vacuous (tautologous, if you wish). Lakatos clearly meant 'unrefuted' in his sense, the correct one.

If we amend the requirement to include only the well-corroborated content of the defeated theory we clash with Lakatos' second way of reducing the conventionalist element in Popper: the appeal procedure by which any bothersome observations (the 'well-corroborated' content of the rival) can be overthrown.[26]

These points aside, the demand would make some portions of T unassailable by methodological fiat, not only when T is replaced by T′ but forever. When T′ is itself replaced that portion of T it had preserved must in turn be preserved by the new successful alternative, and so on. For what other grounds could there be for rejecting it on future occasions of change? And if there are any, why *must* they also be ruled out when T′ replaced T? Thus exposed, the demand is nothing but a methodological disguise for the restriction on radical change discussed in the first part of the chapter. It is worse, in fact, for it assumes that some empirical claims can be established with absolute certainy (otherwise why would anyone dismiss the very possibility of their being overthrown?).

The conservative requirement seemed to ensure a greater degree of continuity in science, and the matter of continuity was thought to have some bearing on the question of the rationality of science. Lakatos' methodology correctly stresses the elaboration of a succession of theories in accordance with the positive heuristic of a program.[27] Science is not a hazardous affair of trial-and-error bush-beating but an intellectually directed enterprise. But rationality lies in how we proceed and change, not in whether our growth is cumulative.[28] Thus the merit of placing metaphysical theories at the core of research programs is not that such a move guarantees the *continuity* of science (in that the resulting theories articulate the metaphysical core), for, as Lakatos realizes, research programs are overthrown by rivals with radically different metaphysical cores.[29] Hence continuity is just a stable period between radical changes. Whether such changes are rational will depend on the manner in which the alternative we choose leads to the growth of our knowledge.

References

1. These concerns bring my *beta'* and gamma' criteria to mind. But then we may begin to see how blurry the distinction is.

2. In his essay, 'Epistemology Naturalized', (from *Ontological Relativity and Other Essays*, Columbia University Press, New York (1969)), Quine distinguishes between the conceptual and the doctrinal sides of epistemology. The doctrinal side deals with the justification of our beliefs about the world, a hopeless enterprise since Hume, according to Quine. The conceptual side attempts to explain how natural knowledge is based on sense experience. After many refinements of Hume's position, the conceptual side of epistemology, and apparently now the heart of the subject, became the attempt to 'account for the external world as a logical construct of sense data', (p. 74). This program amounted to a translation of theoretical terms into observation terms via logic and set theory. Success would have two extraordinarily good consequences: it would make all cognitive discourse as clear as observation terms and logic (and set theory, Quine regretfully adds), and it would 'establish the essential innocence of physical concepts by showing them to be theoretically dispensable', (p. 76). This last matter of justification by elimination seems extremely suspect, but let that pass since, as Quine remarks, the attempts by Carnap and others to carry out the program ended in failure. Proposals that fall short of Carnap's stern rational reconstruction (or rather logical reconstruction) achieve the sort of reduction that does not eliminate, and thus Quine feels that they have no advantage over straight psychology. It seems more sensible to Quine, then, that if we wish for a reconstruction that links science to experience we should settle for psychology. 'Better to discover how science is in fact developed and learned,' he says, 'than to fabricate a fictitious structure to a similar effect.' (p. 78)

I think that Quine's suggestion is indeed very sensible, as far as it goes. But it does not go very far. If one did think that the appropriate pursuits of epistemology were exhausted by the sort of task Carnap had in mind, then Quine's proposal may seem even revolutionary. But why should one think so? Insofar as logical reduction appeared to have some epistemological significance, then replacing it with psychology is a step forward (although, one would hope, not with the black-box psychology suggested by his 'stimulus-meaning' and other proposals). Nonetheless, many other epistemological concerns are not even connected with it. This book is full of them. So are many other books. Something is shared with Quine: a posture, a reaction. But his is very limited.

3. As interpreted by Barry Stroud in 'Conventionalism and the Indeterminacy of Translation', in D. Davidson and J. Hintikka, (eds.), *Words and Objections: Essays on the World of W. V. Quine.* Quine largely agrees with Stroud's interpretation in his reply in the same volume (although he distinguishes between the 'rigidity of logic in translation and the question of the immunity of logic to revision' (p. 317). In other places Quine wavers from this position, e.g., his *Philosophy of Logic*, Prentice-Hall, 1970.

4. Or else it could be 'discovered' that we had three values all along.

Then the radical change would not take place in the conceptual scheme but in our knowledge of it. This sort of move can be made indefinitely to save the claim that only one conceptual scheme is possible, but then such move would be *ad hoc* and unwarranted.

5. Paul K. Feyerabend, 'Consolations for the Specialist', in Imre Lakatos and Alan Musgrave (eds.), *Criticism and the Growth of Knowledge*, Cambridge University Press, London (1972), p. 223.

6. Postulating a primitive symbolic language (e.g., Goodman's proposal in 'The Epistemological Argument', his contribution to The Symposium on Innate Ideas, *Boston Studies in the Philosophy of Science*, Volume III (1968)) would be of very little help in this issue. For starting from such a language what limits can we place on the child's later linguistic abilities? And how can we determine the uniqueness of the symbolic language (or pre-language)?

7. In such a way one could give a liberal interpretation of Bohr's Correspondence Principle in atomic physics. Stronger and more generalized versions of the principle may fall prey to the arguments that follow.

8. Lakatos calls Popper's brand of falsificationism 'naive', (his own version is 'sophisticated', of course).

9. Imre Lakatos, 'Methodology of Scientific Research Programmes', in *Criticism and the Growth of Knowledge*, Imre Lakatos and Alan Musgrave (eds.), Cambridge (1970), p. 116. Future references to this article will appear under the heading *Criticism* (my italics).

10. Imre Lakatos, 'History of Science and its Rational Reconstructions', *Boston Studies in the Philosophy of Science*, Volume 8, R. C. Buck and R. S. Cohen (eds.), p. 100. Future references to this article will appear under the heading *Reconstructions.*

11. *Criticism*, p. 155. Notice the symmetry with the elimination rules for theories.

12. See fns. 9 and 11. Also *Criticism*, p. 118, and *Reconstructions*, p. 100.

13. Lakatos fervently denied this. Feyerabend, on the other hand, holds it. See, for example, chapter 15 of his *Against Method*, NLB, London (1975). (So does Toulmin, cf. his *Human Understanding*, Volume I, Princeton (1972).)

14. Whether the inconsistency is weak or strong will depend on the particular (or the various) interpretation given to the requirement.

15. *Criticism*, p. 119. Lakatos' own italics.

16. In that from the start they are confronted with anomalies. *Criticism*, p. 120 fn. 2, p. 121 fn. 2, p. 124.

17. Within a program the verification of the n + 1th theory amounts to the refutation of the nth. Since many such refutations are foreseen, they play a less crucial role in Lakatos' methodology than in Popper's. *Criticism*, p. 137; *Reconstructions*, p. 101.

18. Can we apply to portions of theories Lakatos' view about theories and research programs? We should be able to. Lakatos himself talks in this manner. His position is clear: '*any part* of the body of science can be replaced but only on the condition that it is replaced by a 'progressive' way'. *Criticism*, p. 187 (my italics).

19. This discussion applies *mutatis mutandis* to research programs.

20. We may still consider T a good theory, compared to a previous theory which it replaced, and thus 'successful', and some of its content 'well corroborated'. But after the defeat of T at the hands of T' we may have a brand new game of corroboration and the like.

21. *Criticism*, pp. 127–31.

22. *Criticism*, pp. 138–40.

23. Brownian motion could not be a 'fact' in the light of an observation

theory partial to the phenomenological program.

24. *Criticism*, p. 157. Curiously enough, history offers a striking counter example to Lakatos' appeal procedure (just an 'anomaly', if you wish): the success of Galileo. His rivals appealed by attacking his observation theory, viz. the theory of the telescope, which was poorly corroborated, inarticulated, and full of 'hidden' assumptions, and which remained poorly corroborated, inarticulated, and full of 'hidden' assumptions long after Galileo's victory (cf. Feyerabend's 'Problems of Empiricism', in Colodny (ed.), *Beyond the Edge of Certainty*).

25. His quoting of Soddy. *Criticism*, p. 140.

26. For Lakatos' discussion of the conventionalist element in his methodology see *Criticism*, pp. 125–31.

27. See Lakatos' discussion of the relative autonomy of theoretical science. *Criticism*, pp. 134–38, and p. 182. Also *Reconstructions*, p. 99.

28. *Criticism*, p. 189.

29. For the role of metaphysics in science, see *Criticism*, pp. 183–84. Some form of continuity is important if Lakatos is correct in emphasizing that we (should) appraise not theories but *series of theories*. We want to know that we are dealing with a *genuine* series and not with the result of *tacking* theory upon theory. In the examples Lakatos considers, the required continuity seems to be provided by apparent preservation of some content, but also by some consistent application of the positive heuristic to the enterprise of theory building. I suggest that the second option is more appropriate if we wish to have a general application of Lakatos' insight. At any rate a distinction must be drawn between continuity within research programs and continuity within *science.*

8

ON CONCEPTUAL SCHEMES

In the preceding chapter I defended the possibility of radical change. This may seem to fly in the face of the analogy with evolution, since the mechanisms of natural selection bring change about very gradually and some (in fact, most) of the previous genetic stock is preserved. It would appear, then, that Lakatos' account would meet the analogy more closely. I can always drop the analogy, of course, but I do not think I have to. I do not have to because, even though the change of our basic intellectual structures must be relatively slow, the change of the behavioral structures and 'conceptual schemes' (which result from the interaction between our basic intellectual structures and the environment) need not be slow.[1]

I make a distinction then between our basic intellectual apparatus and conceptual schemes. Here we may keep in mind the distinction suggested earlier between the intellectual *genotype* and the intellectual *phenotype*. The genotype would correspond to our basic intellectual apparatus, the phenotype to its expression in a variety of milieus (social, cultural, etc.). A conceptual scheme would be a formalization of the phenotype, for (to mimic Strawson *et al.*) we might say that a conceptual scheme is a rendition in logical 'space' of the basic intellectual structures. Even supposing that some conceptual scheme is indeed common to all humans, that is, even if all human thought must conform to a certain pattern, we should distinguish the conceptual scheme from the basic intellectual apparatus, just as we distinguish the spider web patterns from the mechanism that permits spiders to produce them, although it has been claimed that every species has a specific web pattern.

For any conceptual scheme to be as solid as those who trade in such goods require, its corresponding intellectual apparatus should be species specific, very stable, and of a very fundamental nature. Now, the actual organs of thought may already incorporate some options in development, and thus it might be advisable to consider instead the blueprint for such development – hence the title *genotype*. It does not matter for the time being whether we look for such genotype in that group of genes relevant to our intellect or think of it as a theoretical construct. But it does matter whether such intellectual apparatus (strong properties and all) can have only *one* expression. Are we justified in believing that there can be only one conceptual scheme, only one phenotype? No. If radical conceptual change is possible, as argued in the previous chapter, so are a variety of conceptual schemes.[2] I suggest then that *radical* 'conceptual' changes can be better viewed as changes in the phenotype. Now, much of scientific change will take place within a given conceptual scheme, but sometimes it may require radical conceptual revisions. In either case, scientific change can come about much less gradually than that occurring in the intellectual apparatus. Making a fuller use of the analogy with evolution allows us to preserve it: genotypes change slowly, but phenotypes need not.

Some may wish to deny the distinction between our basic intellectual structures and conceptual schemes. For them, conceptual schemes would be as fundamental as anything can be in this matter. The other considerations I have brought up might be interesting, but not from a philosophical point of view. There are several problems with this denial. The first one is that a view of that sort would have great difficulty in accounting for radical change. How a conceptual scheme could be replaced by another must remain completely mysterious. It is in fact beyond the boundaries such an approach sets upon itself: one could only record that a change had taken place. A flexible and dynamic mind would be a pox upon an analytic house. The second problem is that a view of that sort would involve us again in the fastidious separation between conceptual and empirical issues. I will try to show that in this the pure conceptual approach is tainted with empirical assumptions once more.

It will be instructive to consider a positive case for the uniqueness of some conceptual scheme and then see how my claims about conceptual change can be handled in the light of either view of the distinction between basic intellectual structures and conceptual schemes. When taking the first approach (that the distinction is plausible), I will claim that we cannot show that the proposed conceptual scheme is the unique expression of our basic intellectual apparatus. When taking the opposite approach, I will claim that we cannot show that such a scheme is *our* conceptual scheme. I will claim further that the same objections will apply to any other conceptual scheme.

The following, then, is a proposal for finding *our* conceptual scheme (and thus the boundaries of our intellect).

It has been said that a species' science is (at least partially) contingent on its brain (on the central nervous system, or equivalent, of the members of the species). When talking about the brain in the present context, however, we should not limit ourselves to a mass of specialized nerve cells, blood vessels, hemispheres and the like. The main interest should be placed on the *modus operandi* of the human brain. The human brain has certain ways of dealing with the world that are different from those of the frog's or the gorilla's. Speaking loosely we may say that the brain has been *programmed* (a general purpose program, to be sure).[3] A comprehensive psychological theory would involve the spelling out of such a program (although a complete study of man may also require an account of how such a program comes to be realized in the central nervous system. It should be pointed out that nothing rules out in principle the realization of such a program in different materials, i.e., synthetic tissues, electronic circuits, etc.).

At first it sounds plausible to suggest that *if* we were to spell out such a program we would have succeeded in demarcating the limits of human understanding. After all, we would have specified the boundary conditions of the human mind. It may well turn out that there are enormous degrees of freedom, but if there are patterns of thinking, patterns of development and so on, peculiar to the human brain at work, then there are biological (or psycho-biological) 'categories'. How could we tell that we had arrived at such a program? What assurances could we have that our psychological theory was indeed final, complete? The proposal seems to provide a plausible answer to these questions.

Suppose that we were to develop some general purpose programs. We would also, just as we do today, try them out, i.e., bring about their realization, most likely in electronic robots. The robots thus equipped will be able to manage in the environment to different degrees of satisfaction. If our aim were to come up not just with a very good, even an excellent, general purpose program, but with a *human* general purpose program, then we would evaluate the alternative programs based on how closely their total set of responses compare to the total set of responses of a human being. Were a program to enable robots to 'get along' in the universe just as humans do, then we should say of that program that it is *the* program of the human mind.[4]

There could be some minor complications here: not all human beings behave alike, so how could a robot behave *just* like us? Would robots have sexual drives? Would the brightest robot in high school want to take out the prettiest blonde? Would he want to become the captain of the football team? If not, again, how could a robot 'get along' in the world *just* as we do?

There are several ways of avoiding these irritating little questions. We could, as mentioned above, demand not only the *program* but its *realization* in our particular kind of hardware (software?) before we grant the achievement of a complete account of man's nature. That is, we may have reservations about electronic robots, but if using the same program and a good deal of biological engineering we built reassuring *androids* made out of organic tissue (which by then will be living tissue), then we would have to agree that we have constructed human persons. (Some of us may go on refusing to grant them personhood. But then, some of us are prone to irrationality and insane prejudice.) The test of the theoretical pudding, then, is in the *making* (when the result tastes like pudding).

There is no need to go that far. If we agree that we are in possession of the *basic* human program, we would have achieved our goal of psychological Total Knowledge. A basic program would be a *primary* program of *homo sapiens*, as opposed to secondary, specialized, or idiosyncratic smaller programs which are added to the basic one. (Persons may start out the same but they learn different forms of life, Wittgensteinian 'language games'; the more idiosyncratic acquisitions, e.g. wanting to become captain of the football team, would not be part of the *primary* program.)

What is desired is that our robots should not be told apart from human beings in those behavioral respects which are pertinent to intelligence (at any rate, it is the *structure* of our intelligence that we are after).[5] The suggestion made here amounts to a generalized Turing test, just more stringent and of wider scope. It is also compatible with contemporary linguistic philosophy, for we find important philosophers whose positions would fit in nicely with this proposal: Wittengenstein's views on the nature of knowledge and understanding, for example. According to Wittgenstein the test of whether someone understands something is whether he is capable of performing certain tasks, not whether he has the right 'mental' accompaniment. Sometimes, particularly when the performance of the tasks is at an incipient stage, it will be hard to determine whether he is able to carry out the whole task (e.g. completing a rather long series of integers, when only two integers have been given by the subject). But in many cases it will not be at all difficult. The test of whether someone has understanding (at the human level) will not be simply a few, nor many, tasks but rather a network of them. As long as our robots meet such a test we should attribute their program to ourselves.[6]

Now I will direct myself against the main point of the proposal. We should realize that such a general purpose program, such a human program, would be an adaptation to a restricted area of experience. Thus in finding that program we will have only specified a temporary state of equilibrium in the mental evolution of *homo sapiens.* Even though our robots may be equipped with the same program, the question of the construction of a person will remain open; for outside that limited range of experience that determined the criteria for the test the network of responses may be different. To the extent that psychological theories account for similarity of behavior in different informational environments, i.e., those that human beings have not experienced yet, those theories will be successful. But until the appropriate tests are successfully undertaken we should make no claim of completeness. And the problem is that we can never tell whether we have accounted for all the environments possible to us.[7]

To put the point differently, such a program would formalize the *present modus operandi* of the human brain. We would have understood only one stage of a long evolutionary history. In order to give a complete description of all the faculties of the brain we would have to obtain knowledge of the whole history, something we are not in a position to do (there could be, for example, many faculties of the brain that are 'dormant' now, though they may have been active earlier[8] or faculties that have not yet been developed, or triggered, e.g. 'higher' levels of operations in Piaget's view – see below).

To suppose that we had arrived at *the* human program, then, would require several assumptions: that we had complete evolutionary knowledge, that our exposure to the general environment was comprehensive enough, and so on. From previous remarks it can be seen that such assumptions would be partly empirical and by no means plausible. Thus we assail at the same time the purity and the plausibility of the conceptual approach.

The other approach permitted the distinction between conceptual schemes and our basic intellectual structures. In this case the analogy with biology is again useful: in formulating such conceptual schemes we actually formulate our present intellectual *phenotype*. Such an intellectual phenotype would be determined by our basic intellectual structure, the intellectual *genotype*, and a particular environment, or limited range of environments, in the wide sense of 'environment' used at the beginning of the chapter.[9] Given that we are, in Lorenz' phrase, 'specialists in non-specialization', we should be able to exhibit different phenotypes with different exposures to different environments (we may want to reserve the term 'phenotype' for the disjunctive class of the many possible alleged phenotypes, but that would be a most unhelpful attitude, for then we could not determine even the phenotype).[10] Through the favoring of a particular phenotype in a particular environment some environmental pressure may be put on the genotype which, in turn, would favor some of the alternatives in it. But change would come about very slowly. Phenotypes, on the other hand, may come and go more easily and more frequently.

All these considerations may be advanced against any other proposed conceptual scheme, for they do not depend on the circumstances of the particular proposal discussed in this chapter. I conclude, therefore, that claims of uniqueness are not justified with respect to conceptual schemes. No conceptual scheme can be shown to be *our* conceptual scheme.

It is worth remarking, if the analogy with biology is allowed, that the temporary uniqueness of a conceptual scheme is also questionable. It would be rather surprising, for example, that all humans shared *exactly* the same intellectual genetic stock. When talking about genotypes we may grant a good deal of overlap, but also some (if slight) differences in constitution. This should translate into flexibility of phenotype.

Now we may speculate. It could well be (and I hope it is) that *homo sapiens* are capable of much more complex modes of thought (a more advanced 'stage' in Piaget's scheme). It is interesting to note, as Piaget has, that in several primitive societies the adult members never go beyond the level of concrete operations. They fail to reach the level of propositional operations which is attained by adolescents between twelve and fifteen in the developed countries of the West.[11] We can speculate about whether we, in the Western world, might not also be 'primitive' in a similar way (with respect to a future civilization, or even a hypothetical civilization).

With exposure to a different environment (physical, biological, cultural, scientific, technological) we may take the additional steps needed to reach a 'higher' level of operations. Such a new, 'higher' level may again enable us to change our environment so that the new exposure may lead to an even 'higher' level (of which *homo sapiens* are potentially capable now).

It could also be that infants brought up in environments altogether different to any that can be reasonably expected in the near future on this planet (but not so radically different that they would not survive) may undergo a different sequence of levels (particularly as the levels increase in complexity). The accompanying suggestion is that humans may be capable of several sequences of operational levels, some of which may be *superior* to our present one (in that *homo sapiens* so growing up would be able to 'get along better' in the universe).[12]

During the Skylab II mission a most interesting experiment was performed. Spiders are known to weave webs with species-specific patterns (two-dimensional patterns). In order to explore the effect of the lack of gravity on the animal's behavior, a spider was taken aboard the Skylab. For the first twenty-four hours the spider was unable to weave any webs with coherent patterns: it seemed bewildered. Soon enough, however, its webs settled into a specific patterns: a three-dimensional pattern! Given the species-invariant character of the two-dimensional patterns the temptation was there to infer an innate structure isomorphic with web patterns (or at least that the appropriate section of the spider's genotype could find only one phenotypic expression). But the result of the spider's exposure to a non-gravitational environment should do away with the temptation. Our civilized history is rather short; we may be about to enter the pursuit of horizons never dreamed of before. Trying to deal with a very different environment may change our intellectual phenotype in a manner similar to the spider's. (Or we may never be so exposed, but we cannot discount the possibility that if we *were* our conceptual scheme would change radically.)

1. My view is more analogical in some places than in others. Though much can be said for Toulmin's account of Knowledge in a model that mirrors Darwin's theory, I think it places too much distance between science and the world. Such is roughly the point made in chapters 4 – 6. In that instance my position is far more than analogical. In this chapter it is closer to the spirit of Toulmin's. In fact, some of the points I make here are akin to some of those Toulmin makes in his *Human Understanding* (Princeton University Press, 1972). There are some important differences, though. Whereas I am interested in making some remarks of general application, Toulmin aims against very specific views (e.g., the existence of an universal grammar). Toulmin also seems to place great emphasis on the matter of isomorphism, to which I hardly give a passing glance. I find Toulmin's work very impressive, although I cannot say that it influenced the points I make in this chapter (only, perhaps, because I developed them before I read his book, and via a different route). But the telling difference is one of general purpose. Whereas Toulmin searches for a uniformitarian character in the evolution of our scientific disciplines, I am arguing that we ought not to conceive thus of our conceptual schemes, scientific development and the like. Perhaps something more basic intellectually would change only in such uniformitarian manner. That is in fact my suggestion. On the balance, then, we find the possibility of radical change.

2. The possibility in question would be logical or conceptual. The proposal that follows in the text goes beyond that, and thus I think it is a more meritorious way of trying to determine our 'conceptual scheme' than the many 'purely conceptual' attempts that abound in the philosophical literature of the twentieth century. I will not say more on this topic because the battle against Strawson, Quine, *et al.*, was fought in chapter 7. The main function of this chapter is one of clarification.

3. As understood in artificial intelligence and other fields related to computers.

4. The point developed in this proposal is that *if* a certain level of *performance* can be attained by our artificially constructed 'persons', *then* we would be *justified* in claiming that we (and them) share the same program (or conceptual scheme). *The point is not* that general purpose programs can be provided for robots whose brain's hardware would be constituted by *digital* computers. The proposal need not take a stand on this issue. For both sides of the controversy about artificial intelligence the reader may consult the following:

Pro: E. A. Freigenbaum, J. Feldman, (eds.), *Computers and Thought*, McGraw-Hill (1963).
M. Minsky (ed.), *Semantic Information Processing*, MIT Press (1968). Many interesting articles by Simon, Minsky and others can be found in these two anthologies. For general purpose programs applied to language the work of T. Winograd is of particular relevance. A good introduction to his work is 'The Process of Language Understanding', in *The Limits of Human Nature*, J. Benthall (ed.), E. P. Dutton & Co., New York (1974).
Con: Hubert L. Dreyfus, *What Computers Can't Do. A Critique of Artificial Reason*, Harper & Row (1972).

5. This remark is not intended to legislate out of the primary program such things as emotions. It must be obvious that I am being very generous to the proposal. Perhaps all these qualifications are really unnecessary if one talks about behavior falling within a certain range.

6. Such is the proposal for psychological Total Knowledge. It must be noted that if the proposal is correct, there appears to be an important methodological difference between cosmology and psychology. Whereas there would be a *procedure* for settling questions as to whether a particular theory constitutes psychological Total Knowledge, no equivalent procedure would be available for cosmological Total Knowledge. This has a counter-intuitive sound to it, since in dealing with the psychological we have to contend with so many more variables than when dealing with the physical. Counter-intuitive or not, however, we can easily see that no equivalent procedure is open to evaluations of cosmological theories. We can approach the matter at two levels. On the surface, the obvious question is: what would constitute the construction of an universe *just* like ours? It is not just a matter of magnitude. Where would we get the materials? From nowhere? Even if we develop some Hoylean talent for bringing matter and energy into existence, would we be duplicating our universe or just adding to it?

Quite apart from questions of this sort, which may sound silly, we must take into account considerations made in chapter 7.

The possible determination of whether a species had achieved cosmological Total Knowledge was an open question, at least for the species itself. In determining success in the room and explorer cases we made use of a context – a context that was itself part of a larger context, or was connected with other contexts. But it is difficult to see how a similar contextual determination can be made in the case of cosmological Total Knowledge, given its all-embracing character, i.e. it supersedes all other scientific contexts because it is a total theory.

7. The point may come at which talk of exposure to an extended sphere of action seems idle, but I do not think we are close to it yet. When we reach such a point we may or may not have achieved psychological Total Knowledge (after all, we may have simply become over-confident). This situation would parallel that of cosmology. Thus, methodological similarity is assured.

8. Lorenz suggests that the big apes may have shown much more intelligence a long time ago. Due to the lack of use of some of their cognitive faculties, they have become duller. Could they regain their previous intellectual level? Incidentally, Lorenz seems to hold a position similar to mine on the issues discussed in the text.

9. After this discussion we are perhaps in a better position to realize that the intellectual genotype is nothing but a portion of the organism's genome (the ensemble of genes which encodes the structural information for many thousands of proteins). This is not to say that such portion is a subset of the genome 'in charge' of intellect and nothing else, for it is doubtful that sharp boundaries of this sort can be drawn at the level of molecular genetics. Incidentally, at that level we can see, according to Gunther S. Stent, 'that the nature of the relation between that genome and the physical realization of the actual animal, or its phenome, is an extremely complex, and as yet quite unsolved conceptual problem . . . the conceptual obstacle to providing such an account lies mainly in the role played by the enormously complex context in which genes find themselves in the course of embryonic and post-embryonic development.' *Morality as a Biological Phenomenon*, G. S. Stent (ed.), p. 18, Dahlem Workshop Reports, Verlag-Chemie, Life Sciences Research Report 9 (1978).

10. The issue as stated should not be confused with the dispute between genetic and environmental determination. The phenotypes are, after all, expressions of the genotype in different environments. It is important to keep in mind the genotype's function in deciding what environmental factors are relevant to development and how they are relevant. It may be thought fruitful perhaps to

consider the issue in terms of reaction norms. The notion of a reaction norm is not entirely clear, however. In its most appropriate sense to the present discussion, we would understand it to be the collection of traits which 'an organism might develop in all possible sequences and combinations of environments' (David Hull, 'The Trouble with Traits', upcoming in *Theory and Decision*, David Paulsen (ed.)). But I submit that the crucial reaction norms could be determined only for a restricted range of environments.

11. Jean Piaget, *Psychology and Epistemology*, The Viking Press, New York (1972), pp. 61–62.

12. These issues bring up the related issue of whether such stages, as described in Piaget's work, are species invariant. See particularly chapter 3, *ibid.*

9

CONCLUDING REMARKS: THE EPISTEMOLOGY OF THE FUTURE

In this essay I have argued in favor of four main theses: two of substance, we might say, and two of method. The first pair deals with the nature of scientific knowledge. The second divides as follows: one thesis is about philosophical method, which involves us with the proper relationship between science and philosophy; the other is about scientific method, which involves us with the presumably philosophical issue of rationality. I would like to take one final look at all these theses and their interrelationships.

My views on the nature of scientific knowledge have as a foundation two simple assumptions: that the experiences of an organism are the result of an interaction between its cognitive apparatus and the environment, and that the cognitive apparatus itself is the result of a long evolutionary process. From these two claims I developed the Principle of Relativity for epistemology (my first and most startling thesis). The notion at play here is simply that many cognitive 'frames of reference' may be just as 'good' epistemologically, viz. that there are no preferred frames. As a result of the principle, we realized that to know the universe is to know it in some specific way; thus 'reality' is relativized, just as 'mass' was relativized by Einstein's theory of relativity. My view attacks absolutist realism but does not involve us in what has passed for relativism in epistemology. In the first place it insists that the universe be conceived of in a specific way, within a frame of reference; there can be no 'pure' knowledge; for the world to 'be' it must 'be' within a frame of reference in an analogous sense to that in which the 'mass' of an object is its 'mass in a frame of measurement'. There is no *one* way in which the world is. In the second place, my view replaces the *list* model of knowledge with a *performance* model. The distinction between *noumena* and *phenomena* is ruled out without falling into subjectivism: that is the trick that interactionism allows us to perform.

The performance model is the outcome of my second thesis. My approach places me in the company of evolutionary epistemologists, and thus a reminder of the main differences with other approaches might be in order. The account of scientific development, based on an analogy with neo-Darwinism, offers some similarities to other philosophers' positions (Toulmin's, Popper's, and in some respects Feyerabend's). But I go beyond mere analogy, first by developing the *alpha'*, *beta'*, and *gamma'* criteria (an extrapolation to science from Scriven's work on computers), and then by using such criteria in arguing for the importance of 'getting along' in the universe (which sometimes involves what I called 'survival at a distance'). Not to beat around the bush, a consequence of my position is that science has something akin to survival value. This is a bold and risky claim, but I trust that the risk has been worth taking.

These two theses complement each other in the following way. I have argued that there may be no preferred frames of interaction. But I did not want to claim that *all* frames are equally 'good' epistemologically (just that there may be many that are just as 'good'). From this point on the performance model takes its proper place. Among the many intellectual genotypes some will allow for greater degrees of performance in 'getting along' in the universe than in others, and the same can be said of the many phenotypic expressions of any one intellectual genotype.[1]

We need not claim, then, that any all-encompassing theory is as good as any others. Again: there may be different levels of performance. But can there be different global theories such that we should assign the same performance 'grade' to them?[2] Conceivably – the congnitive genotype could be thought of as the ideal limiting case for a particular species. A global theory that exploited all the resources of the genotype would amount to the closest the species could approach to Total Knowledge. It would also constitute the most advantageous phenotype. Before arriving at that point, all sorts of theories may come and go, and in some cases some of them may be equally 'good'. Equality of performance 'grade' with respect to a particular environment is only part of the story, however. There are considerations about the likelihood of enlarging the 'niche', dealing with a changing environment, etc. Those theories are also likely to be just stages of sequences of theories (along the lines of Lakatos' research programs, see chapter 7, thus allowing for Lakatosian considerations about progression and degeneration).

From the two main theses on the nature of knowledge results a most important position on the *limits* of knowledge: the boundaries of our intellect cannot be determined either conceptually or methodologically (the failure of the main attempts drawn from contemporary philosophy was examined in chapter 7). In chapter 8 a comparison was made between radical change and change in intellectual phenotypes. In this connection we should realize that the intellectual genotype need not be rigid ('monolithic' may be an apt expression here), and that it is likely to change, if ever so gradually. Thus conceptual variability is likely to be even greater. It should become clear that an evolutionary approach not only tends not to buttress the status quo, but provides the epistemological underpinnings for the very opposite.[3]

This position on the limits of knowledge is not a consequence of my main epistemological views alone, but also of a particular methodological outlook. My first thesis on method constitutes an attack upon the alleged separation between science and philosophy. Throughout this essay I have talked about the separation between the 'philosophical', or 'conceptual', and the 'empirical' in problems, methodology, etc. Presumably scientists should take care of their problems in a 'scientific' way, the philosophers of theirs in a 'philosophical' way. I have tried to show that crucial, and implausible, assumptions are often made in what philosophers think are purely conceptual investigations. Philosophy should be 'tested' by science; this essay is a step in that direction.

It is important to realize that if we insist on this distinction between science and philosophy, much of the intellectual activity of mankind ends up in limbo. Scientific theories do not come ready made: they originate as vague, unclear stirrings. They must undergo several stages of articulation, clarification, and so on, before they are empirical enough to be called scientific (before, in my account, they meet the *alpha'*, *beta'*, or *gamma'* criteria). The process for empirical readiness may take a very long time (thousands of years in the case of atomism). Often such intellectual stirrings are properly considered philosophical, even metaphysical.[4] The problems that arise in connection with such philosophical concerns should also be considered philosophical. But then, if we are to follow the lead of those who insist on the distinction, we should say that they are linguistic problems! Or worse (according to an important branch of analytic philosophy), they would turn out to be linguistic *mistakes*! If we claim that they are so (and what a ridiculous thing to claim), then it seems that making linguistic mistakes often leads (in the long run) to the growth of knowledge. Truly radical scientific alternatives puncture the conceptual scheme, thus we should not find it strange that the philosophical speculations from which they spring can be at odds with it. That is, they will not conform to the rules of axiomatization employed by the philosophers of the time. We should realize that established scientific theories often become the pillars of our conceptual schemes, e.g. Newtonian physics during the past few centuries, and long before it, Ptolemaic astronomy and Aristotelian science. Even when the casual order is not quite as I describe it, the relationship is often close. Thus it is not surprising that revolutionary theories appear highly implausible at first, e.g. Galileo's case for the Copernican view, and Einstein's theory of relativity. At any rate, it does not seem that such philosophical concerns are amenable to the methodology(ies) of analytic philosophy. Furthermore, many scientific concerns merit the title of'conceptual' even though they are empirically 'connected' (many of those that fulfill the *beta'* criterion, for example).

None of the foregoing should be taken to imply that philosophy becomes a chapter of some part of science or vice versa. Even though the line of separation between the two is often blurred, they are different things. Nor should we be misled by talk of a continuum, with philosophy at one end and science at the other. The relationship between the two is complex and dynamic: they interact at many different points, and when they do they often have (or ought to have) a role in the transformation of the other.

In this context it may be useful to point out what I think is a crucial shortcoming of most conceptual philosophies: what they strive for is the 'mapping out of our concepts', conceptual taxonomy in other words. But taxonomy is at best a very primitive enterprise, and often a very pointless one. Baconian empiricism glorified such taxonomical endeavors in science, conceptual philosophy is its counterpart in professional philosophy. Science undertaken according to the canons of such primitive empiricism is often just a disaster (cf. Feyerabend's remarks on the early history of the Royal Society).[5] Why should we expect taxonomy to take us far in philosophy? Taxonomy is useful when integrated into a general theory that tells us, to begin with, what features are relevant, what ditinctions important, etc. Furthermore, conceptual taxonomy normally aims at the description of a *static* system of knowledge. Thus it can hardly do justice to a dynamic epistemological enterprise. Conceptual investigations can be fruitful only when they take their proper place in a more insightful philosophical approach.

From philosophical method we pass on to the question of scientific method. The world of philosophy of science has reacted almost with consternation to Kuhn and Feyerabend's arguments against method. Method provided us with standards for evaluation. But if Kuhn and Feyerabend's claims concerning scientific revolutions were correct, if change were as radical as they insist it can be, then the standards themselves are subject to change. Without common standards, however, the story goes, there is no way to tell that the changes have led to progress. That is, we cannot be assured of the rationality of science.

This situation must be viewed in the context of a strong tendency to think that it is rational to believe in certain things and irrational to believe in certain others. The lack of method would further aggravate the present epistemological crisis, then, by taking away our means of keeping ill-reputed views in their proper place. If anything goes, rationality goes.

If lacking general recipes for discovery or justification implies that we have no method, then I must agree that we have no method. Certain methodological procedures may be used as they befit the particular situation. But no justificational device, in the normal sense, can be used *tout court*. In addition, I have found in the very nature of scientific knowledge the required reasons for insisting on the possibility of radical change. Nonetheless I feel no need to weep for the departure of rationality, for rationality consists neither in the possibility of evaluation by common standards nor in collecting the 'right' set of beliefs. Rationality should be properly ascribed, if at all, to the enterprise of science as a whole. According to my conception of rationality, if the practice of science is set up in such a way that it not only permits but promotes 'getting along' in the universe, then it is rational. The fulfillment of the conditions that lead to the growth of knowledge, as described in chapter 4, also prove advantageous to an intelligent species. That is why rationality would belong to the *structure* of the scientific enterprise. And, as we have seen, the first demand we must make of such enterprise is that it guarantee the possibility of change. Furthermore, a science with the capability of *radical* change is in an even better position to deal with a recalcitrant universe. Thus radical change, instead of being the nemesis of rationality, would go a long way toward ensuring its success. The tables are turned on the traditional and contemporary epistemologies of science! [6]

Such are the four main theses of *Radical Knowledge*. But 'radical knowledge' is far more than an attack upon the static models of most philosophy of science. It does not, for example, limit itself to an introduction of *history* into our epistemological deliberations (as the evolutionary epistemologists and some more recent philosophers have done). Radical knowledge owes its birth to the contemplation of *future* possibilities. I have not been so concerned with how my account fits the history of science (important as it is), but rather with how our epistemology may fit what is to come. By looking mainly forward we could establish the connection between knowledge and action. But is it not the establishing of such connection that places philosophy back in the center of the pursuit of wisdom?

Our civilized existence has so far been rather parochial, but science has put us on the verge of new horizons. Radical knowledge can open new avenues full of bewildering possibilities. If change is in the offing, if we are to confront radically different forms of life and intelligence (some of which may be of our own creation, at least initially), there is a need for an epistemology that can begin to take such situations into account. An epistemology of that sort, an epistemology of the future (or perhaps, *for* the future), will cease lagging behind science: it will join science as we attempt to abandon the primitive ways of our intellect.

References

1. There is no guarantee, of course, that we will be able to evaluate the performance potential of any cognitive frame, particularly when we are confronted with cases of radical differences. Incidentally, the evaluation itself is carried out *from* a particular frame, thus it surely is a relative notion as well.
2. That is, can my Principle of Relativity be applied in intra-species disputes? Of course. In this connection Feyerabend once suggested that such a principle is implicitly recognized in the epistemological foundation of the complementary principle in quantum physics, in which two radically different ways of looking at the world are fused in harmoniuous pragmatism.
3. I believe this lesson extends to sociobiology, where it has been ignored by some over-enthusiastic practitioners. I refer in particular to the lack of caution in making inferences from phenotypic expressions (e.g., cultural traits)

to genotype.

4. We need not restrict ourselves to tried and true examples like atomism and evolution: consider, say, the notion of the universal matrix of the Bootstrap hypothesis in particle physics. See, for example, Geoffrey F. Chew, 'Bootstrap: A Scientific Idea? ', *Science*, 161 and 'Hadron Boostrap: Triumph or Frustration? ', *Physics Today*, October 1970.

5. See, for example, his 'Explanation, Reduction, and Empiricism', in *Minnesota Studies in the Philosophy of Science*, Volume 3, 1962, Feigl, Maxwell (eds.), or Kuhn's in his *The Structure of Scientific Revolutions*, University of Chicago Press, 1970.

6. The individual is free to follow his instincts, his dreams, or his visions in what may appear promising to him. Constant failure may dim the promise in some cases. In other cases we are confronted with disciplines (e.g., astrology) which by all lights not only have a bad track record but also fail to provide even general hints as to how the tide may turn in their favor. This sort of discipline is hard to take seriously.

The quest for certainty, and then for method, imposed on us a mistaken epistemological model. I hope this sketch will serve to rectify matters in part. We should not fear uncertainty. Nor should we feel that it is somehow unphilosophical. For out of our uncertainty came progress and adventure. Why must we require more? After all, should not philosophy be born out of wonder?

INDEX